ZOOPLANKTON OF THE GREAT LAKES

ZOOPLANKTON OF THE GREAT LAKES

A Guide to the

Identification and Ecology

of the Common

Crustacean Species

MARY D. BALCER

NANCY L. KORDA

STANLEY I. DODSON

THE
UNIVERSITY
OF
WISCONSIN
PRESS

Published 1984

The University of Wisconsin Press
114 North Murray Street
Madison, Wisconsin 53715

The University of Wisconsin Press, Ltd.
1 Gower Street
London WC1E 6HA, England

Work on this book was funded in part by the University of Wisconsin Sea Grant College Program under a grant from the National Sea Grant College Program, National Oceanic and Atmospheric Administration, U.S. Department of Commerce, and by the State of Wisconsin (federal grant number NA80AA-D-00086, project number R/LR-12). The U.S. government is authorized to produce and distribute reprints for government purposes notwithstanding any copyright notation that may appear hereon.

First printing

Printed in the United States of America

For LC CIP information see the colophon

ISBN 0-299-09820-6

Plates drawn by Nancy L. Korda

Figures 3b, 17–19, 22, 24–26, 28–33, 38–43, 45–50, 55, 71, 77, and 78 are redrawn after the second edition of Ward and Whipple's *Freshwater Biology*, edited by W. T. Edmondson, published by John Wiley and Sons, Inc., 1959, and used here with their kind permission.

Figures 73, 83, 84, 88, and 90–93 are from K. Smith and C. H. Fernando, 1978, *A guide to the freshwater calanoid and cyclopoid copepod Crustacea of Ontario*, University of Waterloo Biology Series 18.

Figures 23 and 27 are redrawn after C. L. Herrick and C. H. Turner, 1895, *Synopsis of the Entomostraca of Minnesota*, in *2nd Report of State Zoologist*.

Contents

Life History and Ecology of the Major Crustacean Species

Notes on the Distribution and Abundance of the Less Common Crustacean Zooplankton

List of Plates

Acknowledgments

We would like to thank the following individuals and public agencies for their assistance in sampling the zooplankton of Lake Superior: Gary Curtis and Jim Selgeby of the U.S. Fish and Wildlife Service in Ashland, WI; Gary Fahnenstiel and Dr. Robert Keen of Michigan Technological University-Houghton; Mort Purvis of the Ontario Ministry of Natural Resources; Arthur Lasanen, captain of the fishing vessel *Atomic*; Dr. William Swenson, Dave Anderson, Sue McDonald, Sue Medjo, Lynn Goodwin, Lisa Schmidt, Gloria Berg, and Tom Markee of the University of Wisconsin-Superior's Center for Lake Superior Environmental Studies; Lt. Commander Lundberg of the U.S. Coast Guard Cutter *Mesquite* and Frank Johnson of Sivertson's Fishery Company.

Additional samples of zooplankton species from Lake Erie were provided by Roberta Cap of the Great Lakes Laboratory, Buffalo, NY.

Assistance in analyzing the zooplankton samples was provided by Tim Linley and Carolyn Lie.

Sections of this manuscript were edited by E. Bousfield, S. Czaika, J. Gannon, J. Havel, S. Gresens, D. Krueger, C. Lie, and A. Robertson.

We would especially like to thank Linda McConnell, Grace Krewson, Denise Rall, Jacque Rust, and Doris Brezinski for typing several drafts of this manuscript and Cheryle Hughes for preparing several of the illustrations used in the taxonomic key.

ZOOPLANKTON OF THE GREAT LAKES

Introduction

When studying an aquatic system, it is necessary to identify the individual components of the ecosystem and examine their relationships. In the Great Lakes, the crustacean zooplankton play an important role in the transfer of energy from the primary producers, the algae, to the higher order consumers such as aquatic insects, larval fish, and some adult fish. Any disturbance, such as nutrient enrichment, fish introductions, thermal discharges, or toxic effluents, that alters the composition of the zooplankton community could ultimately affect the rest of the system.

This taxonomic key and accompanying information were prepared to facilitate the identification and study of crustacean zooplankton communities of the Great Lakes. The key is designed so that students and researchers with only a limited knowledge of crustacean taxonomy can learn to identify the common Great Lakes zooplankton, primarily using a dissecting microscope. The life history and ecology section summarizes information on the major crustacean species to help readers determine the ecological role of each species.

At the present time, approximately 100 species of crustacean zooplankton have been reported from the Great Lakes. Most of these organisms are restricted to littoral or benthic habitats and seldom occur in plankton collections. In this report, we emphasized the life history and ecology of 34 species most frequently collected from the nearshore and limnetic environments. The key covers 42 Great Lakes species and references other published information to identify the remaining species.

The taxonomic key is based largely on zooplankton collected from Lake Superior. Samples of some Great Lakes species not found in Lake Superior were obtained from other investigators. We examined several adult males and females of each of the common species to determine species-specific characteristics that are readily visible at 50× magnification, then used these characteristics to construct a dichotomous key. Although the key is primarily concerned with the common species, you can identify most organisms to family or genus. Dissections and greater magnification are often needed to identify the less common species. In these cases, you are directed to other works with more detailed species descriptions. To assist in the identification process, detailed composite drawings of the dominant zooplankton species found in the Great Lakes are also provided (see plates).

The scientific names used here are consistent with those in the 2nd edition of Ward and Whipple's *Freshwater Biology* (Edmondson 1959a), except in cases of changes since its publication. The more current names are referenced in our list of "Species of Crustacean Zooplankton Found in the Great Lakes" and in the taxonomic history sections for individual species.

Common names of fish are in accordance with the American Fisheries Society (1980).

Information on the distribution, abundance, life histories and ecology of the Great Lakes crustacean zooplankton was obtained through an extensive literature search. This information was supplemented by data collected during a year-round sampling program conducted on Lake Superior.

A 0.5-meter diameter, 80 μm mesh net was towed vertically to obtain plankton samples from several nearshore and pelagic stations located in the United States waters of Lake Superior. Samples were collected weekly in the Duluth-Superior region during the ice-free seasons of 1978–1980 and at monthly intervals during the winter. Monthly samples were also collected at the Sault Ste. Marie outfall of the Great Lakes Power Co. and around the Keweenaw Peninsula during 1979 and 1980. Other stations in the western half of the lake were sampled once a year during the summer months.

All samples were preserved in 4% formalin and returned to the lab for analysis.

General Morphology
and Ecology
of the Crustacean
Zooplankton

CLADOCERANS

ANATOMY

Cladocerans, commonly known as water fleas, are generally 0.2–3.0 mm long. The body is not distinctly segmented and is enclosed in a folded shell-like structure, the carapace, that opens ventrally (Fig. 1). The shape of the carapace varies. In some species it terminates in a spinule or spine, referred to as a mucro in the Bosminidae. In *Leptodora* and *Polyphemus* the carapace is greatly reduced and restricted to the brood chamber.

The head, which contains a single, darkly pigmented compound eye, is not enclosed by the carapace but is protected by a separate, hardened head shield. In some species a tiny pigmented light-sensitive organ, the ocellus, is located below the eye. Small, often inconspicuous antennules (first antennae) are attached to the ventral surface of the head. The antennules often have olfactory setae. Projecting from the sides of the head are the large antennae (second antennae), which are the principal appendages used in swimming. Most cladocerans have segmented, biramous antennae with a variable number of setae or swimming hairs located on each branch. At the base of each antenna, the head shield is modified into a strengthening ridge, or fornix.

In many cladocerans a pointed beak, or rostrum, projects from the head near the antennules. The anterior portion of the head may also be elongated to form a helmet.

Cladoceran mouthparts are small and difficult to see. They are generally located slightly posterior to the junction of the head and body. The darkly pigmented mandibles, which are used to crush and grind food particles, can be seen in constant motion in live animals. Ingested food moves into a tubular intestine that passes through the body and terminates at the anus located on the postabdomen. The intestine of some species is looped or convoluted, while in others it contains blind pouches, or caeca, to aid in digestion by increasing gut passage time.

The central portion of the body, the thorax, bears 4 to 6 pairs of flattened legs covered with finely spaced setae. In some species the first two pairs of legs may be modified to aid in scraping or clinging to vegetation. In *Leptodora* and *Polyphemus* the thoracic legs are cylindrical, which enables these animals to feed raptorially.

Cladocerans carry their eggs in a brood chamber, a space located between the body wall and the dorsal surface of the carapace. Developing embryos are held in this chamber by the fingerlike abdominal processes. Occasionally, embryos may be released from the brood chambers of preserved animals. The embryos do not have well-developed carapaces or appendages (Fig. 4) and should not be confused with free-living juveniles and adults.

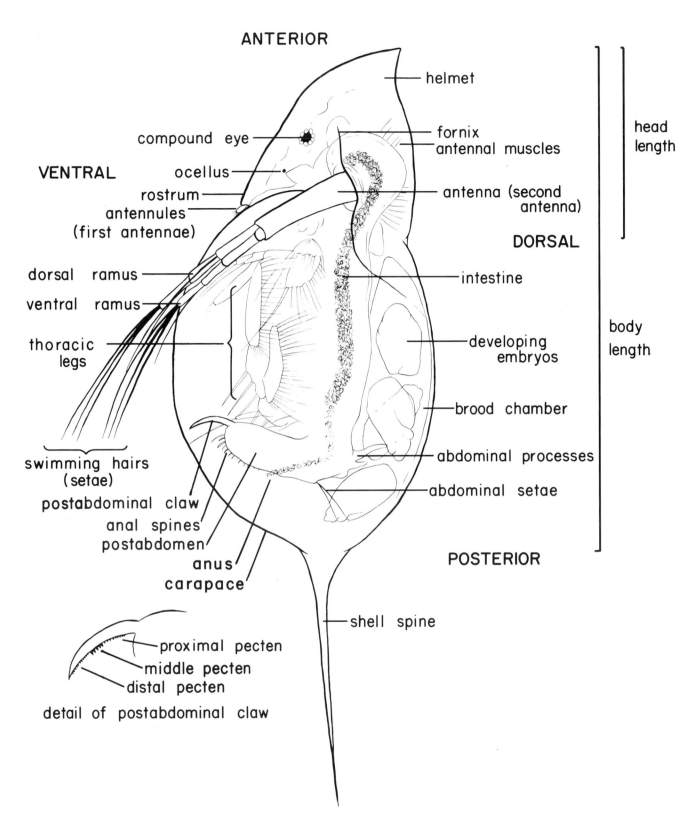

ANTERIOR

helmet

fornix
antennal muscles

compound eye

VENTRAL

ocellus

rostrum

antennules
(first antennae)

antenna (second
antenna)

DORSAL

dorsal ramus

intestine

ventral ramus

thoracic
legs

developing
embryos

brood chamber

abdominal processes

swimming hairs
(setae)

postabdominal claw

anal spines

postabdomen

anus

carapace

abdominal setae

POSTERIOR

body
length

head
length

shell spine

proximal pecten

middle pecten

distal pecten

detail of postabdominal claw

Figure 1   Morphology of a generalized cladoceran

The most posterior region of the body, the postabdomen, terminates in two hooklike cuticular claws. In many species the shape of the fine teeth or pecten located on the postabdominal claws is used in species identification. The teeth of the pecten often differ in size in the proximal, middle, and distal portions of the claw. The postabdomen may also bear rows of spines that aid in removing debris from the thoracic legs.

Male cladocerans (Fig. 58) are usually smaller than females. They have longer antennules and modified postabdomen; some may also possess stout, hooklike claspers on the first legs.

FEEDING BEHAVIOR

Most cladocerans are primarily filter feeders. Movement of the thoracic legs creates water currents that bring suspended food particles into the area between the valves of the carapace. The setae on the thoracic legs strain the particles out of the water and move them towards the mouth. Algae, protozoa, bacteria, and organic detritus of a suitable size are ingested. Large particles and undesirable material can be removed from the filtering appendages and rejected by the modified first legs or by the postabdomen with its terminal claw and lateral spines.

The Chydoridae and Macrothricidae are adapted for feeding in vegetated areas or among the bottom detritus. Their modified first and second legs enable these animals to cling to and scrape food particles off of the vegetation and detritus.

Polyphemus and *Leptodora* possess cylindrical prehensile limbs that allow them to grasp larger prey including protozoans, rotifers, small cladocerans, and copepods.

GROWTH AND REPRODUCTION

In order to increase in size, cladocerans must shed their exoskeletons. After each molt, the animal takes up water and rapidly increases in volume before the newly developed exoskeleton can harden. Cladocerans molt from two to five times before reaching maturity and 6 to 25 times after.

During most of the year, cladocerans reproduce asexually (parthenogenetically). After each molt, mature females deposit 2 to 20 eggs in the dorsal brood chamber. The eggs develop without fertilization into juvenile females that look like miniature adults. At the next molt, the parent females release the free-swimming juveniles, which begin feeding, grow, and mature quickly. This parthenogenetic pattern of reproduction permits a rapid increase in population size under favorable conditions.

When conditions become less favorable due to over-crowding, accumulation of metabolic wastes, decreasing food availability, decreasing water temperatures, or changes in light intensity, some females begin to produce parthenogenetic eggs that hatch into males instead of females. If the adverse conditions continue, some mature females then produce one or two sexual or resting eggs. After copulating with a male, the female releases the fertilized eggs into a specialized brood chamber. The walls of this brood chamber are thick and dark-colored, forming a saddlelike case, known as an ephippium, around the eggs. When the female molts, the ephippium and its enclosed eggs are released; they may sink to the bottom, attach to the substrate, or float at the surface. Ephippial eggs are resistant to freezing and drying. When favorable conditions return, the eggs hatch into females that begin to reproduce parthenogenetically. Most cladoceran populations overwinter as ephippial eggs.

CYCLOMORPHOSIS

Many cladocerans show changes in their morphology as the population develops through the year. This cyclomorphosis may affect helmet shape, eye size, and length of antennules or shell spines. For example, in *Daphnia*, many populations have short, round helmets during the late fall, winter, and early spring. As the population begins to increase in the spring, individuals with more elongated helmets appear (Fig. 59). The helmets of each succeeding generation become increasingly larger until late summer. The head shape of each generation then begins to revert to the rounded form. The causes of cyclomorphosis are not well understood. Water temperature, turbulence, variations in light intensity, genetic variability, and selective vertebrate and invertebrate predation may all be involved in this phenomenon.

VERTICAL MIGRATION

Many cladocerans undergo diurnal vertical changes in position in the water column. Most populations tend to concentrate at the surface at dusk and then move downward again at dawn. Some species rise and sink twice during the night, while others display a pattern of reverse migration, with the greatest surface concentration occurring at dawn. Vertical movement varies from about 1 to 25 m for freshwater cladocerans.

Although changing light intensity appears to be a cue for movement, the migration patterns are also affected by age and size of the animals, food supply, day length, oxygen concentration, turbulence, and several other factors. The adaptive value of this diurnal movement is not yet well understood, but it may be involved in increasing metabolic efficiency and avoiding predators.

More information on the ecology and life histories of cladocerans can be found in Hutchinson (1967), Pennak (1978), and Kerfoot (1980).

COPEPODS

ANATOMY

Adult copepods, commonly known as oarsmen, of the suborders Calanoida, Cyclopoida, and Harpacticoida are 0.3–3.2 mm long. Their elongate cylindrical bodies are clearly segmented (Fig. 2).

The head segments are fused and covered by the carapace. A single, small, pigmented eyespot is usually present along with several paired appendages. The uniramous first antennae (antennules) consist of 8 to 25 segments used in locomotion and chemo- and mechanoreception. The right antennule of adult male calanoids is geniculate (bent) and may possess a lateral projection on the antepenultimate (second from the last) segment (Fig. 2f). Adult male cyclopoids (Fig. 2b) and harpacticoids have both antennules geniculate. The second antennae (Fig. 2c) are generally shorter than the antennules. They are uniramous in cyclopoids, biramous in calanoids, and prehensile in harpacticoids. Copepod mouthparts consist of the paired mandibles, maxillules, and maxillae.

The first thoracic segment is usually fused to the head. This fused body region is termed the cephalic segment and it bears the maxillipeds which are used for feeding. The second through sixth thoracic segments each bear one of the five pairs of swimming legs. Some of these segments may be fused to the cephalic segment or to each other. The first four pairs of thoracic legs are biramous and similar in appearance (Fig. 2d), the inner branch known as the endopod and the outer branch termed the exopod. The fifth pair of legs is usually quite different from the preceding pairs. It is greatly reduced in both sexes of the cyclopoids and harpacticoids. In adult female calanoids, the symmetrical fifth legs are usually modified. They are often reduced in size (Fig. 2h) and may be uniramous or absent in some species. Adult male calanoids generally have asymmetrical fifth legs, and in diaptomids the right exopod terminates in a hooklike claw (Fig. 2g). The shape of this claw and the position of the lateral spine located on the last segment of the exopod are useful in determining species.

The genital segment of adult copepods is often larger than the following abdominal segments. In cyclopoids and harpacticoids, the genital segment may bear a pair of vestigial sixth legs that are more developed in males than females.

The abdominal segments lack jointed appendages. The first abdominal segment may fuse with the genital segment and is followed by one to four additional distinct segments. The last segment bears two cylindrical caudal rami that terminate in hairlike caudal setae. The number, position, and length of these setae (Fig. 2a) are useful in species identifications.

The cephalic and thoracic segments are commonly referred to as the metasome; the genital segment and the abdominal segments make up the urosome. The body of calanoid copepods is constricted between the metasome and the urosome while the cyclopoid and harpacticoid body is constricted between the segments bearing the fourth and fifth legs. In the calanoids and cyclopoids, the segments posterior to the constriction are much narrower than the anterior segments (Figs. 2a and e), but in harpacticoids (Fig. 2i) the urosome is not noticeably narrower than the metasome. The last segment of the metasome of calanoids may be expanded laterally into metasomal wings.

FEEDING BEHAVIOR

Calanoid copepods are primarily filter feeders (Richman et al. 1980). Movements of the second antennae and mouthparts create water currents that carry food particles past the sensory first antennae and into the reach of the feeding appendages. The setae on the maxillae are used to filter these particles out of the water. The feeding current is maintained during feeding, but filtration probably occurs only when a food particle is within reach of the mouthparts. Recent studies have shown that calanoids are capable of selectively filtering certain sizes and types of algae. Some species also use their modified maxillipeds to grasp algae or smaller zooplankton.

Cyclopoids lack developed filtering setae but have mouthparts modified for grasping and chewing. These raptorial cyclopoids may be herbivores, omnivores, or carnivores with different species showing preferences for detritus, algae, protozoans, cladocerans, or other copepods.

Harpacticoids are benthic organisms. They crawl about the bottom and use their modified mouthparts to select edible particles out of the detritus.

GROWTH AND REPRODUCTION

Reproduction in all copepods is sexual. The male clasps the female with his modified first antennae and/or fifth legs, then transfers a packet of sperm, the spermatophore, from his genital pore to her genital segment. The sperm are

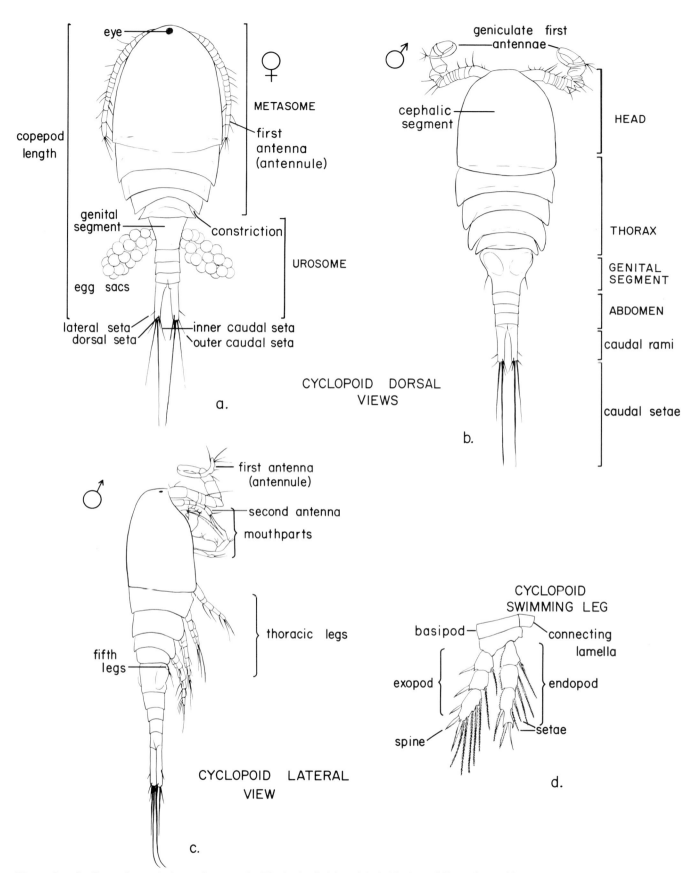

Figure 2a–d General morphology of copepods. The body divisions labeled in 2a and 2b apply to either sex.

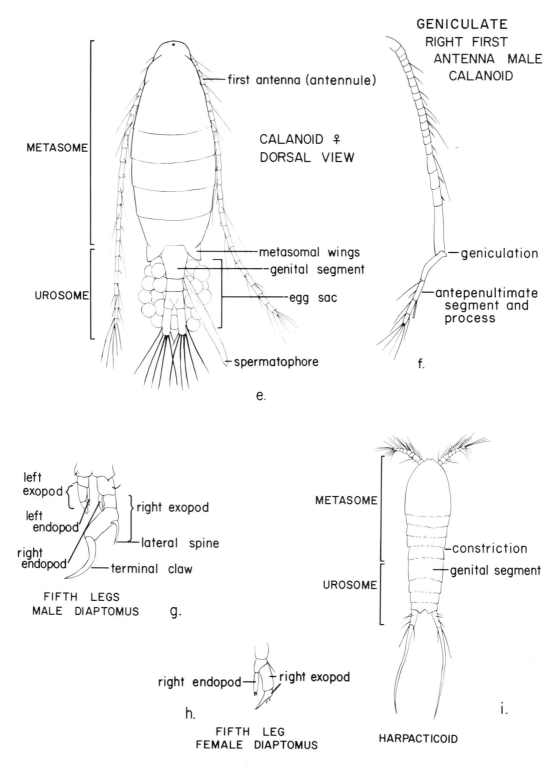

first antenna (antennule)

CALANOID ♀
DORSAL VIEW

METASOME

UROSOME

metasomal wings
genital segment
egg sac

spermatophore

e.

GENICULATE
RIGHT FIRST
ANTENNA MALE
CALANOID

geniculation

antepenultimate
segment and
process

f.

left exopod
left endopod
right endopod

right exopod
lateral spine
terminal claw

FIFTH LEGS
MALE DIAPTOMUS   g.

right endopod — right exopod

h.

FIFTH LEG
FEMALE DIAPTOMUS

METASOME

UROSOME

constriction
genital segment

i.

HARPACTICOID

Figure 2e–i General morphology of copepods

stored in a seminal receptacle located in the female's genital segment. When the female releases eggs from her genital tract, they are fertilized by the stored sperm. Females of most species brood the eggs in one (calanoids) or two (cyclopoids and harpacticoids) egg sacs attached to the genital segment. A few species of calanoid copepods do not brood their fertilized eggs but release them directly into the water, where they sink to the bottom before hatching.

After the first clutch of eggs has hatched, some females fertilize a second and third clutch utilizing more of the sperm stored from their first mating.

Copepod eggs hatch into small, active larvae known as nauplii (Fig. 5). The first nauplius stage (NI) is characterized by three pairs of appendages. The animals grow rapidly and molt to the second nauplius stage (NII), which is slightly longer and, in cyclopoids, possesses a rudimentary fourth appendage. The animals continue to grow and add appendages as they pass through six naupliar stages. The next molt is to the first copepodid stage (CI). At this point the young copepods have the general body shape of the adult but are smaller and lack several of the swimming legs (Fig. 6). Growth, addition of swimming legs, and modification of the limbs continue at each molt until the adult (CVI) stage is reached. The animals then mate and produce the next generation.

The length of copepod life cycles and the number of egg clutches produced by females are quite variable. Some species grow and mature rapidly, producing several generations each year. Others require up to a year to reach maturity. Growth is generally slowed in cold temperatures. Some copepods become dormant and diapause near the bottom during the winter, while others diapause during warm water conditions.

ECOLOGY

Copepods are more common and show a greater diversity in marine systems than in freshwater. In limnetic regions of the Great Lakes, one or two calanoids and one cyclopoid generally dominate. In the littoral areas where there is a greater variety of habitats, more copepod species are encountered. Although some cyclopoids are found among the bottom debris of the littoral and benthic zones, these areas are usually occupied by harpacticoids.

Some copepods are strong swimmers and may undergo diurnal vertical migrations up to 100 m. As in the cladocerans, the stimuli and adaptive significance of this movement are not yet well understood.

MALACOSTRACANS

Order Mysidacea

ANATOMY

Adult mysids, commonly known as opossum shrimp, are 15–25 mm long and superficially resemble shrimp or crayfish. A thin carapace covers most of the thorax but is not fused posteriorly to the thoracic segments (Fig. 3a).

The large compound eyes are located on stalks. Two pairs of antennae also extend from the head; the first pair is biramous while the second pair is uniramous with an elongated basal projection, or scale.

The head contains the paired mandibles, first maxillae, and second maxillae, which are used in food handling. The first two pairs of thoracic appendages, the maxillipeds, are modified for straining zooplankton, phytoplankton, and detritus from the water and moving the particles forward to the mouth. The first pair of maxillipeds is also involved in respiration. Mysids lack gills and respire through the thin lining of the carapace. The first maxillipeds direct a current of water under the carapace to increase the potential for gas exchange.

The posterior six pairs of thoracic appendages are known as swimming legs, or pereiopods. The outer branch (exopod) of each leg projects to the side of the body and is responsible for the smooth, rapid swimming pattern of the animal. The inner branch (endopod) of each pereiopod is used to create a water current that brings food particles to the maxillipeds.

The abdomen of mysids consists of six segments. The first five pairs of abdominal appendages are often reduced in size and are known as pleopods. The last abdominal segment terminates in a flattened telson and a pair of uropods. An equilibrium organ, the statocyst, is located at the base of each uropod.

Mature male mysids are distinguished by their greatly elongated fourth pleopods and the shape of the third pleopods, which are modified for copulation (Fig. 8). Adult females have a set of plates known as oostegites projecting ventrally from the last two thoracic segments. These plates enclose the ventral brood chamber. Juvenile mysids resemble the adults but do not have fully developed reproductive appendages as mentioned above.

Only one member of the Order Mysidacea is common in the Great Lakes. The life history, feeding behavior, and ecology of *Mysis relicta* are described in more detail in a later section.

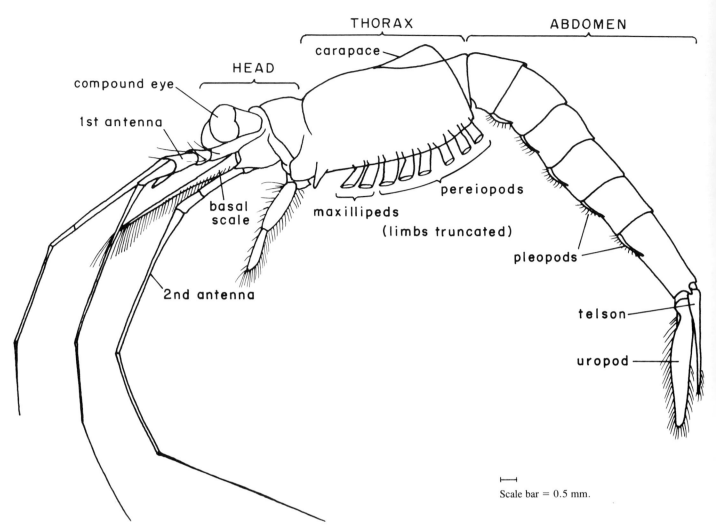

THORAX **ABDOMEN**

carapace

HEAD

compound eye

1st antenna

basal
scale

pereiopods

maxillipeds

(limbs truncated)

2nd antenna

pleopods

telson

uropod

⊢——⊣
Scale bar = 0.5 mm.

Figure 3a General morphology of a malacostracan, Order Mysidacea, lateral view

Order Isopoda

ANATOMY

The aquatic pill bugs or sow bugs of the order Isopoda are not generally planktonic but are occasionally found in samples collected near breakwalls or from the bottom of vegetated areas of the Great Lakes.

The 5–20 mm long adults are dorsoventrally flattened and lack a carapace (Fig. 3b). The head possesses two pairs of antennae; the second pair longer than the first. The compound eyes are not stalked and are located on the dorsal surface of the head. The remaining ventrally located head appendages—the mandibles, first maxillae, and second maxillae—are used for feeding.

The first thoracic segment is fused to the head and has paired, flattened ventral appendages, the maxillipeds, which are used in food handling. The next seven thoracic segments are expanded laterally. Each possesses a pair of uniramous walking legs, the pereiopods. The first pair, gnathopods, may be modified for grasping and are more specialized in males than in females.

The first two abdominal segments are reduced in size; the succeeding four segments are larger and fused together. The modified appendages, the pleopods, of the first five segments are hidden beneath the abdomen and aid in respiration. In males, the second pair of pleopods are modified into copulatory stylets and are used for sperm transfer. The last abdominal segment bears the paired uropods, which project posteriorly from the body.

FEEDING BEHAVIOR

Isopods are scavengers and crawl along the bottom mainly in shallow water searching under rocks and among vegeta-

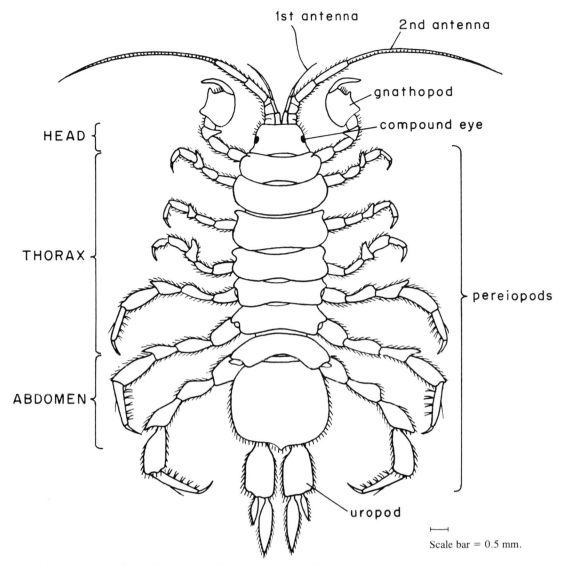

1st antenna

2nd antenna

gnathopod

compound eye

HEAD

THORAX

ABDOMEN

pereiopods

uropod

Scale bar = 0.5 mm.

Figure 3b General morphology of a malacostracan, Order Isopoda, dorsal view

tion and debris for dead and injured animals. They also consume leaves, grass, and aquatic vegetation.

The ecology and life history of the isopods are described in more detail by Pennak (1978).

Order Amphipoda

ANATOMY

Commonly known as scuds (suborder Gammaridea) these animals typically lack a carapace, and the body segments are laterally compressed. Scuds from the Great Lakes are 5–20 mm long when mature. Bousfield (1982) is an excellent source for the anatomy and taxonomy of amphipods.

The head (to which the first thoracic segment is fused) has paired, unstalked, sessile compound eyes located laterally (Fig. 3c). Two pairs of antennae are present. The first antenna consists of a three-segmented basal penduncular portion and a longer, more slender multisegmented flagellum; a shorter accessory flagellum is present on Great Lakes scuds. The second antenna is longer, with a peduncle of 5 segments, the second bearing the excretory gland cone. The mouthparts are clustered in a compact mass directly beneath the head. They consist of the unpaired and fused upper and lower lips, the paired movable mandibles, first and second maxillae, and the maxillipeds, the latter pair fused basally.

The seven free segments of the thorax each bear a pair of seven-segmented uniramous "walking" legs, pereiopods. The first two pairs are usually modified for grasping, often

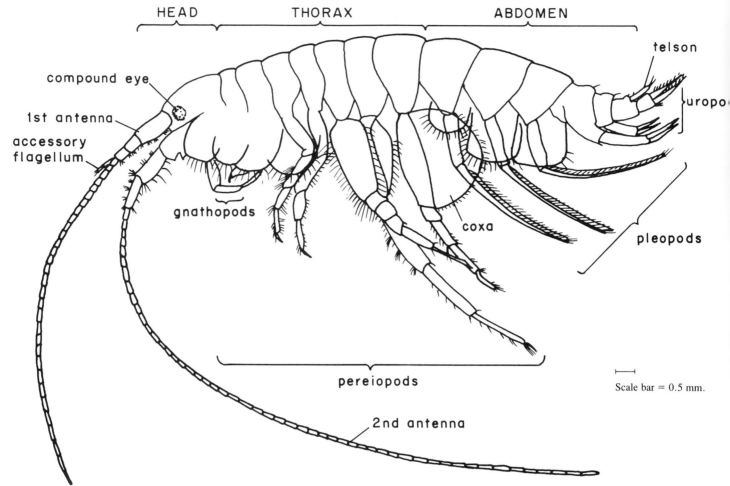

HEAD   THORAX   ABDOMEN

telson

uropo

compound eye

1st antenna

accessory
flagellum

gnathopods

coxa

pleopods

pereiopods

Scale bar = 0.5 mm.

2nd antenna

Figure 3c   General morphology of a malacostracan, Order Amphipoda, lateral view

strongly so in the male, and are termed gnathopods. The basal segment, coxa, of each leg is enlarged and flattened into an outer protective plate. Attached to the inner surface of coxae 2–7 (often lacking on 7) are flattened respiratory sacs, coxal gills. Slender, fingerlike accessory gills are attached to the sternum in some groups, including *Pontoporeia* and *Hyallela*.

In mature females (Fig. 9), the coxae of pereiopods 2–5 bear large, thin interlocking brood plates (oostegites) that form a ventral brood chamber. In mature males, the short, paired penis papillae are located near the coxae of pereiopod 7.

The abdomen consists of six segments and terminates in a short, usually bilobate or platelike telson. The first three segments have paired biramous flexibly multisegmented pleopods used in swimming and maintaining a respiratory current. The posterior three abdominal segments (occasionally fused together) bear stiff biramous uropods, which aid in pushing, swimming, and burrowing.

FEEDING BEHAVIOR

Amphipods are voracious feeders but only a few groups are predaceous on live animals. When consuming detritus or plant material, they use the gnathopods to grasp and hold the material while the mouthparts chew off portions. Some amphipods are filter feeders, and some, mostly marine, are external parasites or commensals on fishes or colonial invertebrates.

GROWTH AND REPRODUCTION

Many species of scuds become sexually mature at about the 9th or 10th molt stage or instar. In gammarids, a sexually mature male often pairs with a subadult (8th instar) female and clings to her for 1–7 days until she molts. The animals then mate. The female extrudes the eggs into her ventral brood chamber where they are fertilized by sperm strings from the male. Eggs are incubated for 1–3 weeks. The newly hatched young may be retained in the chamber for an additional 1–8 days, then released when the female molts. Adults molt every 3–40 days and may molt several times after they reach sexual maturity.

It is difficult to determine the sex of the very immature instars. Sixth or seventh instar females begin to form small oostegites, and developing eggs may be observed in the ovaries. In some amphipods (such as *Gammarus* and *Hyallela*) males are distinguished by their modified, enlarged gnathopods, which begin to differentiate from the female form at about the sixth instar. All later immature males may be identified by the small penis papillae at the bases of the 7th pereiopods.

ECOLOGY

Most scuds are generally benthic organisms with a negative response to light. They hide among rocks, vegetation, and detritus in shallow water. Pennak (1978) and Bousfield (1973) give more information on the ecology of the amphipods.

General Procedures for Collecting and Identifying Crustacean Zooplankton

Zooplankton may be collected with water samplers, plankton traps, plankton nets, Clarke-Bumpus samplers, and pumps. Edmondson and Winberg (1971) describe each of these methods and discuss their advantages and disadvantages. In the Great Lakes, most investigators collect zooplankton by taking a vertical tow with a 0.5-m diameter plankton net. Samples are usually concentrated and preserved in a solution of 4% neutralized formaldehyde (equals 10% formalin solution).

Subsamples are counted when the total number of organisms in a sample is too large to be counted easily. In these cases, measure the sample volume and thoroughly mix it before using a wide mouth pipette to remove a known volume of the sample. Place the subsample in a Sedgewick-Rafter cell, a plankton counting cell, or a small gridded Petri dish for examination.

Most organisms can be identified to major taxonomic group and the common plankton can be identified to species with a dissecting microscope (25–50× magnification) and the following key. Dissections and higher magnification (100–200×) are needed to confirm the identity of some organisms.

Cladocerans are commonly identified from the lateral view (Fig. 1). Use small needles or insect pins mounted on wooden dowels or inserted into the eraser of a pencil to manipulate the animals into the proper position. In some cladocerans you must examine the pecten of the postabdominal claw to confirm the identification. In this case, place the organism on a glass slide in a small drop of water. Use the probes to reach into the carapace, pull the postabdomen out, and cut it off. Then place a cover glass over the postabdomen and observe it under a compound microscope.

Copepods are often identified from the dorsal view. Probes must be used to hold the animals since the position of the legs tends to roll them over onto their sides. Note that legs are omitted or drawn in an ideal location in the following illustrations of copepods. You must examine the fifth leg of some copepods to identify them correctly. In calanoids the fifth legs can be examined by pulling them to the side of the animal (Figs. 79–82) or cutting them off (Fig. 83). The fifth legs of cyclopoid copepods (Fig. 2c) are very tiny and are best observed by cutting the animal in half between the segments with the fourth and fifth legs and mounting the posterior half, ventral side up, on a slide.

When examining a sample of zooplankton, it is a good idea to look through the sample, key out several of the organisms, and examine them carefully. When you become familiar with the organisms, you can then begin counting the sample by moving the dish or slide slowly, without splashing, and identifying the organisms you see in each grid. More information on subsampling and counting samples is provided by Edmondson and Winberg (1971). It is a good idea to preserve some of the organisms in small vials or to make permanent slides of them (see Edmondson 1959b) so that the identification can be confirmed.

The following key identifies the common Great Lakes zooplankton to species. Many of the less common animals are keyed only to family or genus because accurate species identification would require dissections and use of high magnification. While figures of representative organisms in the family or genus are provided, we caution you against identifying an animal solely on the basis of the figure and advise you to consult the more detailed taxonomic works listed.

DISTINGUISHING LIFE STAGES

This key is applicable for all free-living life stages of the common Great Lakes cladocerans. Except where the key refers specifically to males or females, either sex can be used to identify a species. Occasionally, embryos (Fig. 4) will be released from the brood chambers of preserved females. These immature forms lack developed appendages and carapaces and cannot be identified correctly by the following key; they should not be counted as free-living organisms.

Characteristics of adult copepods (stage VI copepodids: CVI) are used for species identification in this key. Although they resemble adults, immature copepodids often lack the necessary characteristics for identification and should be separated from the adults before applying the key. The copepod nauplii are generally small (100–400 μm

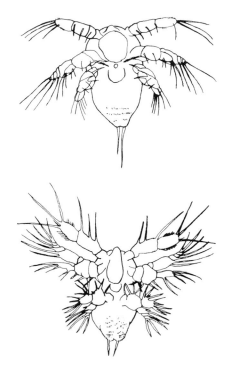

Figure 5 Copepod nauplii

long) and resemble the animals shown in Fig. 5. The body shape and number of appendages vary in the different species and naupliar stages.

Adult (CVI) male cyclopoids are recognized by their geniculate first antennae (Fig. 2b) while adult females usually are found carrying paired egg sacs. Adults of both sexes have an enlarged genital segment that is longer than any of the posterior abdominal segments. The number of abdominal segments depends on the developmental stage and sex of the animal. Adult males have four segments posterior to the genital segment while adult females have only three. In immature cyclopoid copepodids (CI–CV), the genital segment is similar in size to the succeeding segment, and the terminal abdominal segment is the longest since it hasn't yet divided transversely (Fig. 6).

Adult male calanoids have only the right first antenna geniculate (Fig. 2f). In preserved specimens the first antennae may be extended, but the geniculation (joint) is still evident (Figs. 80–81). CVI males have five abdominal segments; fewer segments are present in stages CI–CV. Adult females are more difficult to distinguish from immature stages since the number of adult abdominal segments varies between species. They are most easily recognized when carrying eggs or spermatophores. The genital segment is usually expanded slightly and, in a lateral view, is seen

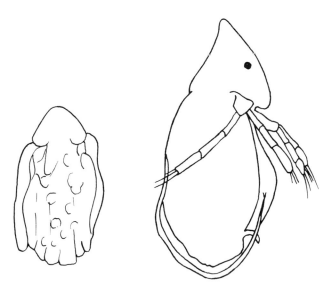

Figure 4 Cladoceran (*Daphnia*) embryos

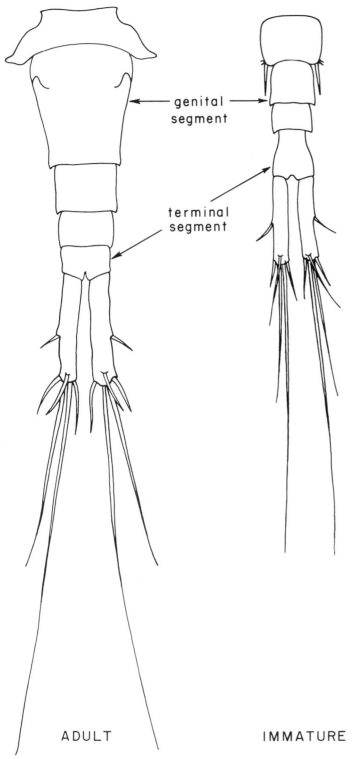

genital
segment

terminal
segment

ADULT   IMMATURE

Figure 6 Abdomen of cyclopoid copepods (*Diacyclops*), dorsal view

to bulge ventrally (Figs. 61, 63). Immature calanoid copepodids are smaller than the adults and lack the above characteristics.

Detailed keys to the nauplii (NI–NVI) and copepodids (CI–CVI) of all common Great Lakes species of cyclopoids and calanoids (Czaika 1982) and to the immature diaptomid copepodids (Czaika and Robertson, 1968) are available but require dissection of appendages and high magnification. Torke (1974) describes the naupliar and copepodid stages of *Limnocalanus macrurus*, diaptomid spp., and *Diacyclops thomasi*, and Katona (1971) describes the immature stages of *Eurytemora affinis*, but neither provides species keys.

USING A TAXONOMIC KEY

A taxonomic key provides a simplified method to differentiate organisms from each other. Most keys are arranged in couplets consisting of a pair of phrases that give alternative choices; i.e., an organism either does or does not have a certain characteristic. To use the following taxonomic key to the Great Lakes crustacean zooplankton, begin at couplet 1. Decide whether the organism you want to identify is best described by phrase 1a or 1b. Pointers on figures point out major features described by the phrases. Then go to the couplet indicated by the number to the far right of the phrase you selected. Again determine which phrase of the couplet best describes your unknown organism. Continue to move to the couplets indicated by your choice of phrases until you come to the name of an organism or group of organisms. Read the additional descriptive notes that may be present and compare your unknown animal with the figures indicated. If they agree, you now know the identity of the animal. If your animal does not resemble the figure and description, you most likely selected the wrong phrase in one of the preceding couplets. Go back and decide where you may have been mistaken and then continue along from the alternative phrase until you learn the identity of your unknown animal. You may also wish to consult the more detailed descriptions and life history information provided for the major species after the key. Page numbers for this information are provided in the key at the point of identification for the species covered.

Note: This key is designed for the more common species of crustacean zooplankton. Some rare species will not be accurately described by any of the couplets and will not resemble the accompanying figures. In these cases more detailed keys should be consulted. Appendages have been omitted from some of the figures in order to show underlying body features.

Classification of Great Lakes Crustacean Zooplankton

PHYLUM ARTHROPODA
Class Crustacea
SUBCLASS BRANCHIOPODA
ORDER CLADOCERA
Suborder Haplopoda
Family Leptodoridae—*Leptodora*
Suborder Eucladocera
SUPERFAMILY POLYPHEMOIDEA
Family Polyphemidae—*Polyphemus*
SUPERFAMILY SIDOIDEA
Family Sididae—*Diaphanosoma, Latona, Latonopsis, Sida*
Family Holopedidae—*Holopedium*
SUPERFAMILY CHYDOROIDEA
Family Daphnidae—*Ceriodaphnia, Daphnia, Scapholeberis, Simocephalus*
Family Moinidae—*Moina*
Family Bosminidae—*Bosmina, Eubosmina*
Family Macrothricidae—*Acantholeberis, Drepanothrix, Ilyocryptus, Macrothrix, Ophryoxus, Wlassicsia*
Family Chydoridae—*Acroperus, Alona, Alonella, Alonopsis, Anchistropus, Camptocercus, Chydorus, Disparalona, Eurycercus, Graptoleberis,*

Kurzia, Leydigia, Monospilus, Pleuroxus, Rhynchotalona

SUBCLASS COPEPODA
ORDER EUCOPEPODA
Suborder Calanoida
Family Pseudocalanidae—*Senecella*
Family Centropagidae—*Limnocalanus, Osphranticum*
Family Temoridae—*Epischura, Eurytemora*
Family Diaptomidae—*Leptodiaptomus, Skistodiaptomus*
Suborder Cyclopoida
Family Cyclopidae—*Acanthocyclops, Cyclops, Diacyclops, Eucyclops, Macrocyclops, Mesocyclops, Paracyclops, Tropocyclops*
Suborder Harpacticoida
Family Canthocamptidae—*Bryocamptus, Canthocamptus, Epactophanes, Mesochra, Moraria*
Family Ameiridae—*Nitocra*

SUBCLASS MALACOSTRACA
ORDER MYSIDACEA
Family Mysidae—*Mysis*
ORDER AMPHIPODA
Family Pontoporeiidae—*Pontoporeia*
ORDER ISOPODA

Species of

Crustacean Zooplankton

Found in

the Great Lakes

Species Name[a]	Identification Level[b]	Page Reference	Location[c]				
			Erie	Huron	Michigan	Ontario	Superior
CLADOCERANS							
Acantholeberis curvirostris (O. F. Müller) 1776	F	32, 112		x			
Acroperus harpae Baird 1843	F	28, 112	x		x		x
Alona affinis (Leydig) 1860	F	28, 112	x	x	x	x	x
Alona circumfimbriata Megard 1967	F	28, 112			x		
Alona costata Sars 1862	F	28, 112	x		x		x
Alona guttata Sars 1862	F	28, 113	x		x		x
Alona lepida Birge 1893	F	28, 113					[x]
Alona quadrangularis (O. F. Müller) 1785	F	28, 113	x		x		
Alona rectangula Sars 1861	F	28, 113	x		x		
Alonella excisa (Fisher) 1854	F	28, 113	x				
Alonella nana (Baird) 1850	F	28, 113	x				x
Alonopsis elongata Sars 1861	F	28, 113					x
Anchistropus minor Birge 1893	F	28, 113		x	x		x
Bosmina longirostris (O. F. Müller) 1785	S	31, 66	x	x	x	x	x
Camptocercus macrurus (O. F. Müller) 1785	F	28, 113	x				
Camptocercus rectirostris Schødler 1862	F	28, 113	x	x	x	x	
Ceriodaphnia lacustris Birge 1893	G	33, 58	x	x	x	x	
Ceriodaphnia laticaudata P. E. Müller, 1867	G	33, 58	x				
Ceriodaphnia pulchella Sars 1862	G	33, 58	x				
Ceriodaphnia quadrangula (O. F. Müller) 1785	G	33, 58	x	x	x		
Ceriodaphna reticulata (Jurine) 1820	G	33, 58	x				
Chydorus faviformis Birge 1893	F	28, 71		x			
Chydorus gibbus Sars 1890 (*Chydorus gibbus* Lilljeborg 1880)	F	28, 71	x				x

Species Name[a]	Identification Level[b]	Page Reference	Location[c]				
			Erie	Huron	Michigan	Ontario	Superior
CLADOCERANS (*continued*)							
Chydorus globosus Baird 1850	F	28, 71	x	x			x
Chydorus latus Sars 1862	F	28, 71	x				
Chydorus sphaericus (O. F. Müller) 1785	F	28, 71	x	x	x	x	x
Daphnia ambigua Scourfield 1947	S	35, 111	x	x	x		
Daphnia galeata Sars 1864 *mendotae* Birge 1918	S	36, 60	x	x	x	x	x
Daphnia longiremis Sars 1861	S	37, 64	x	x	x	x	
Daphnia parvula Fordyce 1901	S	34, 111			x		
Daphnia pulex Leydig 1860 emend. Richard 1896	S	35, 111	x	x	x	x	x
Daphnia retrocurva Forbes 1882	S	38, 62	x	x	x	x	x
Daphnia schødleri Sars 1862	S	35, 111			x		
Diaphanosoma brachyurum (Liéven) 1848	G	27, 54	x	x	x		x
Diaphanosoma birgei Kořinek 1981 (*D. leuchtenbergianum* Fischer 1850)	G	27, 54	x	x	x	x	x
Disparalona acutirostris (Birge) 1878 (*Alonella acutirostus* (Birge) 1878)	F	28, 113					x
Disparalona rostrata (Koch) 1841 (*Alonella rostrata* (Koch) 1841)	F	28, 113			x		x
Drepanothrix dentata (Eurén) 1861	F	32, 112		x			x
Eubosmina coregoni (Baird) 1850	S	31, 69	x	x	x	x	x
Eubosmina longispina (Leydig) 1860	G	31, 66	[x]	[x]		[x]	
Eurycercus lamellatus (O. F. Müller) 1785	F	28, 113	x	x	x	x	x
Graptoleberis testudinaria (Fischer) 1848	F	28, 114	x	x	x		
Holopedium gibberum Zaddach 1855	S	26, 56	x	x	x	x	x
Ilyocryptus acutifrons Sars 1862	F	32, 112			x		
Ilyocryptus sordidus (Liéven) 1848	F	32, 112	x		x		
Ilyocryptus spinifer Herrick 1884	F	32, 112	x			x	x
Kurzia latissima (Kurz) 1874	F	28, 114	x	x			
Latona setifera (O.F. Müller) 1785	S	27, 110	x	x	x		x
Latonopsis occidentalis Birge 1891	S	28, 110	x				
Leptodora kindti (Focke) 1844	S	25, 49	x	x	x	x	x
Leydigia acanthocercoides (Fischer) 1854	F	28, 114	x				
Leydigia leydigi (Schødler) 1863 (*L. quadrangularis* (Leydig) 1860)	F	28, 114	x	x	x		
Macrothrix laticornis (Jurine) 1820	F	32, 112	x		x	x	x
Moina micrura Kurz 1874	S	33, 111			x		
Monospilus dispar Sars 1861	F	28, 114	x	x			x
Ophryoxus gracilis Sars 1861	F	32, 112		x	x		
Pleuroxus aduncus (Jurine) 1820	F	28, 114	x				
Pleuroxus denticulatus Birge 1878	F	28, 114	x		x		
Pleuroxus hastatus Sars 1862	F	28, 114					x
Pleuroxus procurvus Birge 1878	F	28, 114	x		x		x
Pleuroxus striatus Schødler 1863	F	28, 114	x				
Polyphemus pediculus (Linné) 1761	S	26, 52		x	x	x	x
Rhynchotalona falcata (Sars) 1861	F	28, 114					x
Scapholeberis aurita (Fischer) 1849	G	34, 111	x		x		
Scapholeberis kingi Sars 1903	G	34, 111	x		x		x
Sida crystallina (O.F. Müller) 1875	S	28, 110	x	x	x	x	x
Simocephalus exspinosus (Koch) 1841	G	34, 111					x
Simocephalus serrulatus (Koch) 1841	G	34, 111	x		x		
Simocephalus vetulus Schødleri 1858	G	34, 111	x				x
Wlassicsia kinistinensis Birge 1910	F	32, 112			x		

Species Name[a]	Identification Level[b]	Page Reference	Location[c]				
			Erie	Huron	Michigan	Ontario	Superior
CYCLOPOID COPEPODS							
Acanthocyclops vernalis (Fischer) 1893 (*Cyclops vernalis* Fischer 1893)	S	47, 93	x	x	x	x	x
Cyclops scutifer Sars 1863	S	47, 115		x			
Cyclops strenuus Fischer 1851	S	47, 115					x
Diacyclops nanus (Sars) 1863 (*Cyclops nanus* Sars 1863)	S	46, 115	x	x			
Diacyclops thomasi (S. A. Forbes) 1882 (*Cyclops bicuspidatus thomasi* S. A. Forbes 1882)	S	48, 96	x	x	x	x	x
Eucyclops agilis (Koch) 1838	S	47, 115	x		x		x
Eucyclops prionophorus Kiefer 1931	S	46, 115			x		
Eucyclops speratus (Lilljeborg) 1901	S	47, 115	x				x
Macrocyclops albidus (Jurine) 1820	S	45, 115	x		x		x
Mesocyclops edax (S. A. Forbes) 1891	S	46, 98	x	x	x	x	x
Paracyclops fimbriatus poppei (Rehberg) 1880	S	46, 116	x		x		
Tropocyclops prasinus mexicanus Kiefer 1938	S	46, 101	x	x	x	x	x
CALANOID COPEPODS							
Epischura lacustris S. A. Forbes 1882	S	41, 77	x	x	x	x	x
Eurytemora affinis (Poppe) 1880	S	40, 79	x	x	x	x	x
Leptodiaptomus ashlandi (Marsh) 1893 (*Diaptomus ashlandi* March 1893)	S	42, 44, 82	x	x	x	x	x
Leptodiaptomus minutus (Lilljeborg) 1889 (*Diaptomus minutus* Lilljeborg 1889)	S	43, 44, 84	x	x	x	x	x
Leptodiaptomus sicilis (S. A. Forbes) 1882 (*Diaptomus sicilis* S. A. Forbes 1882)	S	42, 44, 87	x	x	x	x	x
Leptodiaptomus siciloides (Lilljeborg) 1889 (*Diaptomus sicililoides* Lilljeborg 1889)	S	41, 45, 89	x	x	x	x	x
Limnocalanus macrurus Sars 1863	S	38, 74	x	x	x	x	x
Osphranticum labronectum S. A. Forbes 1882	N	114			x		[x]
Senecella calanoides Juday 1923	S	40, 73		x	x		x
Skistodiaptomus oregonensis (Lilljeborg) 1889 (*Diaptomus oregonensis* Lilljeborg 1889)	S	42, 43, 91	x	x	x	x	x
Skistodiaptomus pallidus (Herrick) 1879 (*Diaptomus pallidus* Herrick 1879)	S	42, 43, 115	x			x	
Skistodiaptomus reighardi (March) 1895 (*Diaptomus reighardi* March 1895)	S	42, 45, 115	x	x	x		
HARPACTICOID COPEPODS							
Bryocamptus nivalis (Willey) 1925	SO	38, 116				x	
Bryocamptus zschokkei (Schmeil) 1893	SO	38, 116				x	
Canthocamptus robertcokeri M. S. Wilson 1958	SO	38, 116	x		x	x	
Canthocamptus staphylinoides Pearse 1905	SO	38, 116	x		x	x	
Epactophanes richardi Mrazek 1893	SO	38, 116				x	
Mesochra alaskana M. S. Wilson 1958	SO	38, 116				x	
Moraria cristata Chappuis 1929	SO	38, 116				x	

Species of Crustacean Zooplankton Found in the Great Lakes

Species Name[a]	Identification Level[b]	Page Reference	Location[c]				
			Erie	Huron	Michigan	Ontario	Superior
HARPACTICOID COPEPODS (*continued*)							
Nitocra hibernica (Brady) 1880	SO	38, 116				x	
Nitocra spinipes Bueck 1864	SO	38, 116				x	
MALACOSTRACANS							
Mysis relicta Lovén 1861	S	24, 103	x	x	x	x	x
Pontoporeia Krøyer 1842	G	25, 106	x	x	x	x	x

[a] The names in this list are consistent with those in Ward and Whipple's *Freshwater Biology* (Edmondson 1959a). Where names have changed since its publication, the Edmondson (1959a) name is given in parentheses below the new name.

[b] Identification level

SO = Suborder   F = Family   G = Genus   S = Species   N = Not in Key

[c] Where location is bracketed, the animals given the indicated name were probably misidentified.

Key to Common Great Lakes Crustacean Zooplankton

1a. Two large, paired lateral compound eyes; each body segment with a pair of appendages; large animals (6–25 mm), shrimp- or scudlike (Figs. 3b, 7–10)...... .... Subclass MALACOSTRACA, 2

1b. Compound eye single (median) or absent; appendages lacking on several abdominal segments; small animals (most < 5 mm); waterfleas and oarsmen..... 4

2a. Eyes on stalks; body shrimplike, with carapace (Figs. 7, 8) . . . Order MYSIDACEA, *Mysis relicta*, . . . p. 103

Adults 13–25 mm, young 3 mm.

2b. Eyes not stalked; body lacking carapace 3

3a. Body segments flattened dorsoventrally; abdomen with 5 pairs of pleopods, 1 pair of uropods (Fig. 3b) .... Order ISOPODA

Isopods are rarely planktonic but are common in some bottom samples. Consult keys by Chace et al. (1959) or Pennak (1978) for species identification.

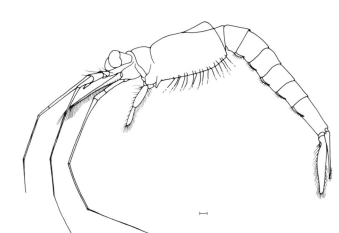

Figure 7 *Mysis relicta*, ♀, lateral view

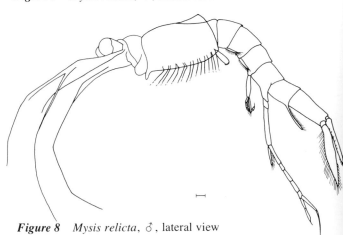

Figure 8 *Mysis relicta*, ♂, lateral view

Scale bars = 0.5 mm.

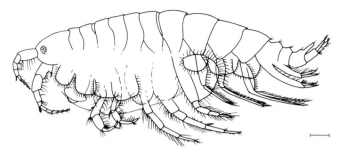

Figure 9 *Pontoporeia*, ♀, lateral view

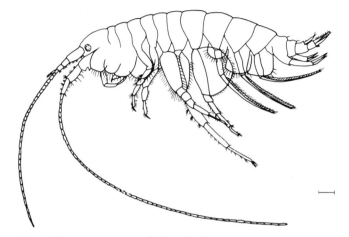

Figure 10 *Pontoporeia*, ♂, lateral view

3b. Body segments compressed laterally; abdomen with 3 pairs of pleopods, 3 pairs of uropods (Figs. 9, 10). Adults 6–9 mm. . . . . Order AMPHIPODA, p. 106

Pontoporeia (pelagic males) occur seasonally (late fall to spring) in the plankton. Chace et al. (1959), Pennak (1978) and Bousfield (1982) describe the other members of this order.

4a. Single compound eye present; longest head appendage (2nd antenna) usually with 2 branches (Figs. 1, 11, 13). . . . .
. . . . Subclass BRANCHIOPODA, ORDER CLADOCERA, 5

4b. Compound eye absent; longest head appendage (1st antenna) with only 1 branch (Fig. 2) . . . .
. . . . Subclass COPEPODA, 26

5a. Body and thoracic appendages enclosed in a shell-like carapace 7

5b. Carapace reduced, only covers brood chamber; appendages exposed 6

6a. Body long and slender (up to 18 mm long); eye small, less than half as wide as head (Figs. 11, 12) . . . .
. . . . *Leptodora kindti*, p. 49

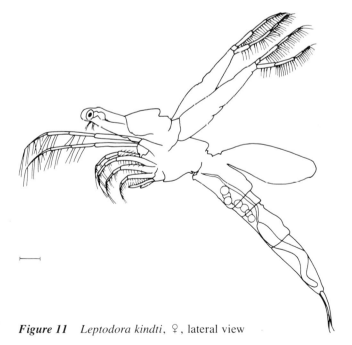

Figure 11 *Leptodora kindti*, ♀, lateral view

Scale bars = 0.5 mm.

6b. Body rounded (< 2 mm long) with elongate "tail"; eye large, almost as wide as head (Fig. 13) . . . . *Polyphemus pediculus*, p. 52

7a. Second antennae of females with 1 branch; a large gelatinous sheath often covers the humpbacked animals (Fig. 14). . . . . *Holopedium gibberum*, p. 56

> Adults 0.5–2.2 mm long. Males are similar to females in shape but have biramous 2nd antennae. The sheath may be lost in preserved animals.

7b. Second antennae with 2 branches; no gelatinous covering; back not humped 8

8a. Many (> 14) setae arranged in a row along one side of the dorsal (longer) branch of the 2nd antennae (Figs. 15–19). . . . . Family SIDIDAE, 9

> Preserved animals often have 2nd antennae extended as in figure 15.

8b. Few (< 10) setae on dorsal branch of 2nd antennae, not arranged in a row along one side of the branch (Figs. 21, 41, 49). 12

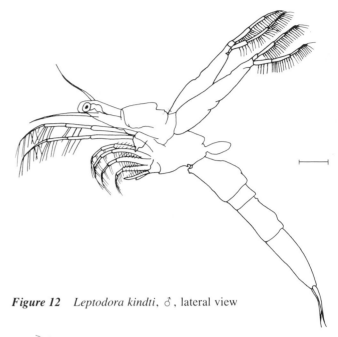

Figure 12   *Leptodora kindti*, ♂, lateral view

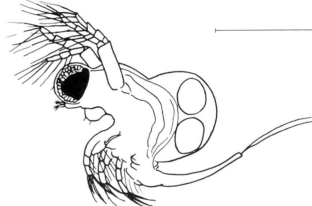

Figure 13   *Polyphemus pediculus*, lateral view

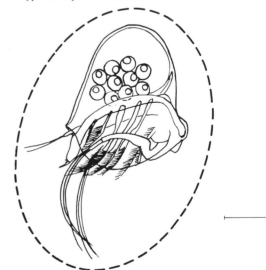

Figure 14   *Holopedium gibberum*, ♀, *lateral view*

Scale bars = 0.5 mm.

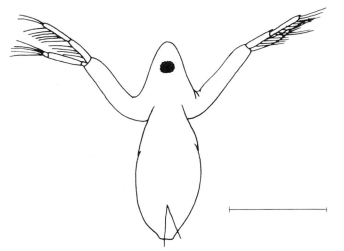

Figure 15 *Diaphanosoma birgei*, ♀, ventral view

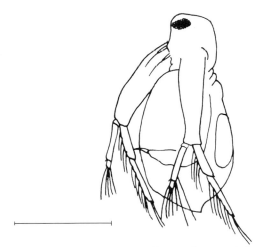

Figure 16 *Diaphanosoma birgei*, ♀, lateral view

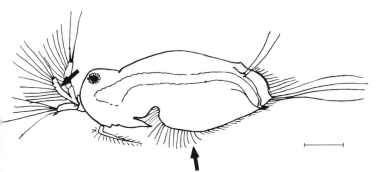

Figure 17 *Latona setifera*, lateral view

9a. Basal segment of 2nd antennae longer than head; no ocellus present (Figs. 15, 16) . . . .
. . . . *Diaphanosoma*, p. 54

Both *D. birgei* and *D. brachyurum* occur in the Great Lakes. See Kořínek (1981) for identification.

9b. Basal segment of 2nd antennae shorter than head; ocellus may be present 10

The following group of littoral sidids is not commonly found in the Great Lakes.

10a. Second antennae with 3 branches, the middle branch an extension from the 1st segment of the dorsal (longest) branch (Fig. 17) . . . . *Latona setifera*, p. 110

Length 2–3 mm; margin of carapace covered with hairs; antennules elongate; ocellus absent.

10b. Second antennae with only 2 branches. 11

Scale bars = 0.5 mm.

11a. Margin of carapace with long setae, some almost ½ height of carapace; ocellus present; length 1.8 mm. two segments in longer (dorsal) branch of 2nd antenna (Fig. 18). . . . . *Latonopsis occidentalis*, p. 110

11b. Margin of carapace with very inconspicuous setae, < 10μm long; no ocellus; length 3–4 mm; three segments in longer branch of 2nd antenna (Fig. 19). . . . . *Sida crystallina*, p. 110

12a. First antennae covered by a beaklike structure; shell spine rarely present (Figs. 20–33). . . . . Family CHYDORIDAE pp. 71, 112

Adult length 0.3–3.0 mm. Second antennae with 3 segments in both branches. *Chydorus* and *Alona* are the most common genera of this family in the Great Lakes, but *Acroperus*, *Alonopsis*, *Alonella*, *Anchistropus*, *Camptocercus*, *Disparalona*, *Eurycercus*, *Kurzia*, *Graptoleberis*, *Leydigia*, *Monospilus*, *Pleuroxus*, and *Rhynchotalona* are occasionally found. Higher magnification and keys by Brooks (1959) should be used to identify members of this family to species.

12b. First antennae exposed, not covered by a beaklike structure (first antennae inconspicuous in some species; Figs. 36, 38, 50); shell spine may be present 13

13a. Paired 1st antennae are fixed, pointed, tusklike structures as long as the head or longer (Figs. 34–37). . . . . Family BOSMINIDAE 14

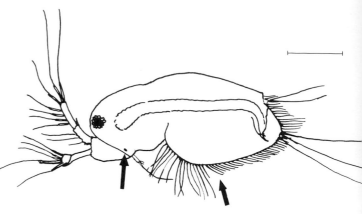

Figure 18   *Latonopsis occidentalis*, lateral view

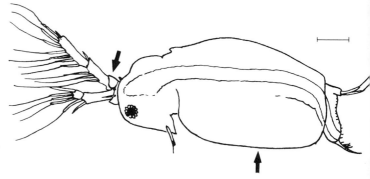

Figure 19   *Sida crystallina*, lateral view

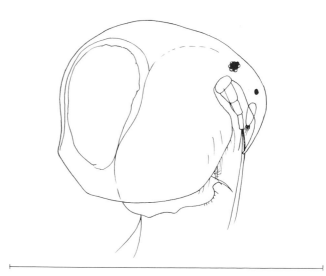

Figure 20   *Chydorus sphaericus*, ♀, lateral view

Scale bars = 0.5 mm.

Key to Common Great Lakes Crustacean Zooplankton

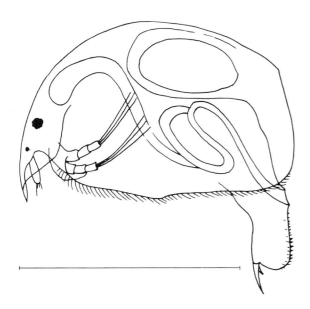

Figure 21   *Alona* sp., ♀, lateral view

Figure 22   *Acroperus harpae*, lateral view

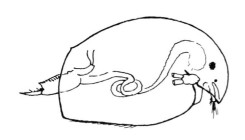

Figure 23   *Alonopsis elongata*, lateral view

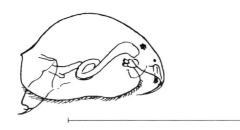

Figure 24   *Alonella* sp., lateral view

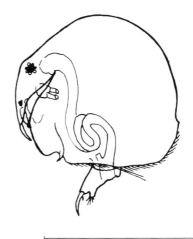

Figure 25   *Anchistropus minor*, lateral view

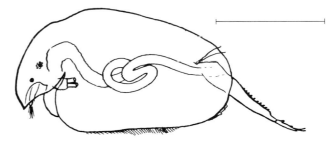

Figure 26   *Camptocercus* sp., lateral view

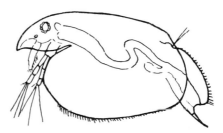

Figure 27   *Eurycercus lamellatus*, lateral view

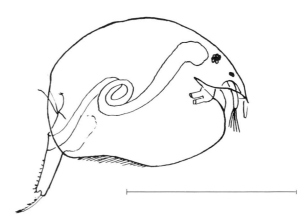

Figure 28   *Kurzia latissima*, lateral view

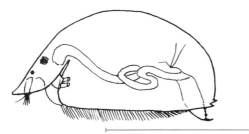

Figure 29 *Graptoleberis testudinaria*,

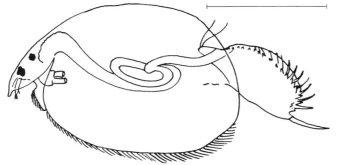

Figure 30 *Leydigia* sp., lateral view

Figure 31 *Monospilus dispar*, lateral view

Figure 32 *Pleuroxus* sp., lateral view

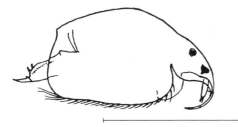

Figure 33 *Rhynchotalona falcata*, lateral view

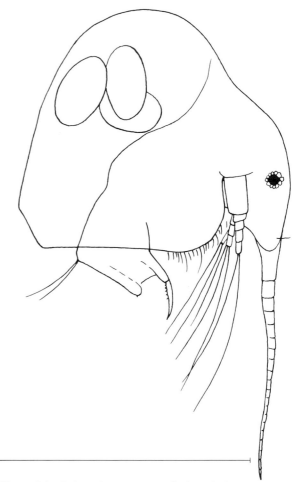

Figure 34 *Eubosmina coregoni*, ♀, lateral view

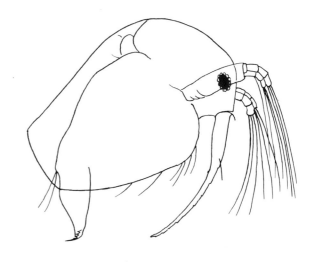

Figure 35 *Eubosmina coregoni*, ♂, lateral view

Key to Common Great Lakes Crustacean Zooplankton

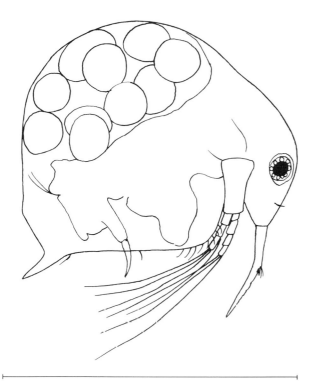

Figure 36 *Bosmina longirostris*, ♀, lateral view

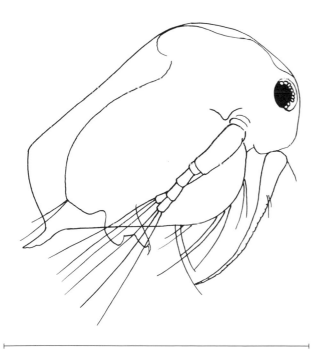

Figure 37 *Bosmina longirostris*, ♂, lateral view

13b. First antennae flexible, tips blunt, bearing setae . . . .
. . . . 15

14a. No shell spine (mucro) present on posterior margin of carapace (Figs. 34, 35). . . . .
. . . . *Eubosmina coregoni*, p. 69

> Length 0.20–0.56 mm. Some populations from Lake Ontario may have a very small shell spine (false mucro). Higher magnification is needed to observe the sensory bristle on the rostrum near the base of the 1st antennae and the pecten located only on the proximal portion of the postabdominal claw. Deevey and Deevey (1971) provide keys to *Eubosmina*.

14b. Shell spine (mucro) present on posterior margin of carapace (Figs. 36, 37). . . . .
. . . . *Bosmina longirostris*, p. 66

> Length 0.4–0.6 mm. Most mucronate bosminids in the Great Lakes are *B. longirostris*. The *Eubosmina longispina* of the older Great Lakes literature is probably a long-featured morph of this species (Torke 1975). Identity can be confirmed by examining, under high magnification, the sensory bristle, located on the rostrum halfway between the eye and the base of the 1st antennae, and the pecten of the postabdominal claw. Both a large proximal pecten and a smaller distal pecten are present. Some mucronate eubosminids with sensory bristles and pecten similar to those of *E. coregoni* may occasionally be found in the Great Lakes. Deevey and Deevey (1971) provide keys to *Bosmina* and *Eubosmina*.

Scale bars = 0.5 mm.

15a. First antennae almost as long as head, attached near front of head (Figs. 38–43) . . . .

. . . . Family MACROTHRICIDAE, p. 112

Length 0.5–2.0 mm. The littoral genera *Acantholeberis*, *Drepanothrix*, *Ilyocryptus*, *Macrothrix*, *Ophryoxus*, and *Wlassicsia* are not commonly found in the Great Lakes. Consult Brooks (1959) for generic and species identifications, which often require higher magnifications.

15b. First antennae variable in length, not attached at front of head 16

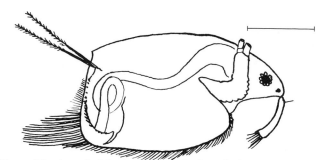

Figure 38   *Acantholeberis curvirostris*, lateral view

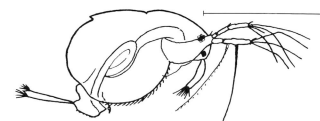

Figure 39   *Drepanothrix dentata*, lateral view

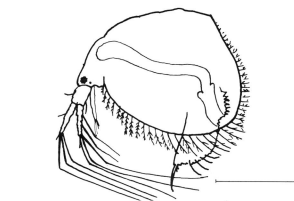

Figure 40   *Ilyocryptus*, lateral view

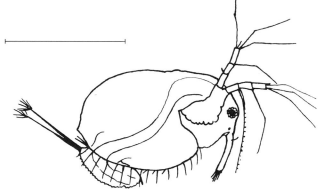

Figure 41   *Macrothrix*, lateral view

Scale bars = 0.5 mm.

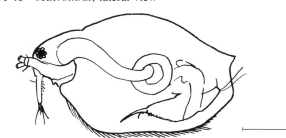

Figure 42   *Ophryoxus*, lateral view

Key to Common Great Lakes Crustacean Zooplankton

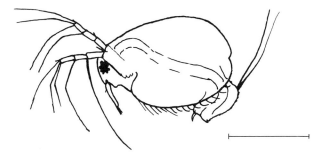

Figure 43 *Wlassicsia kinistinensis*, lateral view

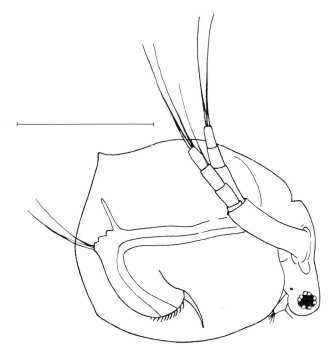

Figure 44 *Ceriodaphnia* sp., lateral view

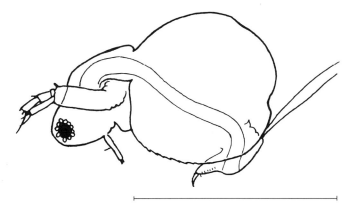

Figure 45 *Moina micrura*, lateral view

16a. Rostrum absent; head curved ventrally (Figs. 44, 45) 17

16b. Rostrum present; head not curved ventrally (Figs. 47, 49). Family DAPHNIDAE (in part) 18

17a. Head small, ⅙ as long as carapace; ocellus present; 1st antennae short, not freely moveable (Fig. 44). . . . . Family DAPHNIDAE, *Ceriodaphnia*, p. 58

Length 0.4–1.4 mm. Five species of *Ceriodaphnia* are found in the Great Lakes. Species identification requires greater magnification to examine features described by Brooks (1959) and Brandlova et al. (1972).

17b. Head large, ½ as long as carapace; ocellus absent; 1st antennae long, flexible (Fig. 45) . . . . Family MOINIDAE, *Moina micrura*, p. 111

Length 0.5 mm; rare in the Great Lakes.

Scale bars = 0.5 mm.

18a. Shell spine present 19

18b. Shell spine absent (Fig. 46). . . *Simocephalus*, p. 111

Length 3–4 mm. *Simocephalus vetulus*, *S. serrulatus*, and *S. exspinosus* are uncommon in the Great Lakes. Species keys using characteristics of the pecten of the postabdominal claw are available in Brooks (1959) and Pennak (1978).

19a. Posterior shell spine located near ventral margin of carapace, ventral margin straight (Fig. 47) . . . .
. . . . *Scapholeberis*, p. 111

Length 1.0 mm. *S. kingi*, with a dark-colored carapace, and *S. aurita*, with a more transparent, white to green carapace, have been reported from the Great Lakes. Brooks (1959) describes both species in greater detail.

19b. Posterior shell spine located near midline of carapace, ventral margin of carapace convex (Figs. 48–59) . . . . *Daphnia*, 20

20a. Shell spine short, < ⅓ carapace length; front of head concave (Figs. 48–50) 21

20b. Shell spine long, > ⅓ carapace length; head shape variable 23

21a. Ocellus present 22

21b. Ocellus absent (Fig. 48). . . *Daphnia parvula*, p. 111

Adults 0.75–1.0 mm. Rostrum short, "pug-nosed." The middle pecten of the postabdominal claw is slightly larger than the proximal pecten (seen only under high magnification). Rare in Great Lakes.

Figure 46 *Simocephalus*, lateral view

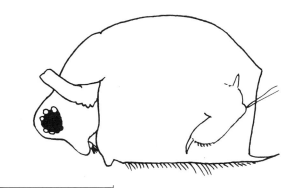

Figure 47 *Scapholeberis*, lateral view

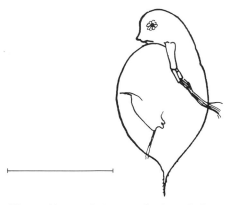

Figure 48 *Daphnia parvula*, lateral view

Scale bars = 0.5 mm.

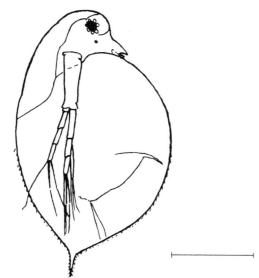

Figure 49 *Daphnia pulex*, lateral view

22a. Large adults, 1.3–2.2 mm long (Fig. 49) . . . .
. . . . *Daphnia pulex*, p. 111

Uncommon in Great Lakes. High magnification shows that the middle pecten of the postabdominal claw is twice as large as the proximal pecten.

22b. Small adults, < 1 mm long (Fig. 50). . . . .
. . . . *Daphnia ambigua*, p. 111

Uncommon in Great Lakes. Pecten on postabdominal claw uniformly fine.

23a. Ocellus present 24

23b. Ocellus absent 25

24a. Head rounded at apex (Fig. 51) . . . .
. . . . *Daphnia schødleri*, p. 111

Adults 1.2–2.0 mm. High magnification reveals enlarged middle pecten of postabdominal claw. Rare in Great Lakes.

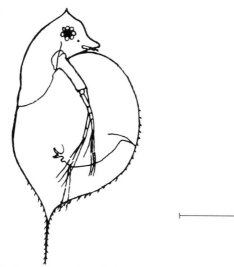

Figure 50 *Daphnia ambigua*, lateral view

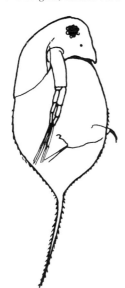

Figure 51 *Daphnia schødleri*, lateral view

Scale bars = 0.5 mm.

24b. Head shape variable, peak located near midline of body (Figs. 52–54). . . . . *Daphnia galeata mendotae*, p. 60

Adults 1.3–3.0 mm. High magnification shows uniformly fine pecten on postabdominal claw. Common in Great Lakes.

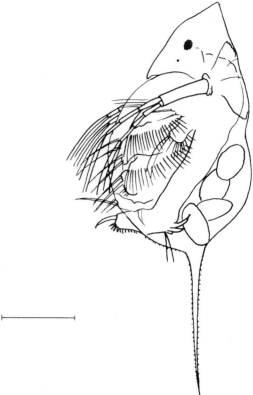

Figure 52 *Daphnia galeata mendotae*, ♀, lateral view

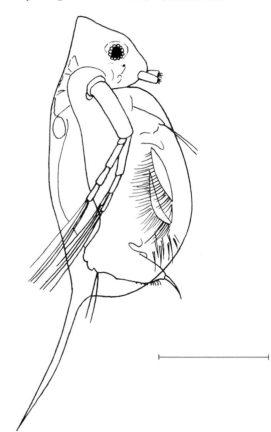

Figure 53 *Daphnia galeata mendotae*, ♂, lateral view

Scale bars = 0.5 mm.

25a. Setae of second antennae reach posterior margin of carapace. (Figs. 55, 56) . . . .
. . . . *Daphnia longiremis*, p. 64

Length 0.8–1.2 mm. Pecten on postabdominal claw unformly small. High magnification also shows that the seta at the base of the second segment of the dorsal ramus of the antennae is shorter than the ramus. Common species.

Figure 54 Variation in helmets of *Daphnia galeata mendotae*

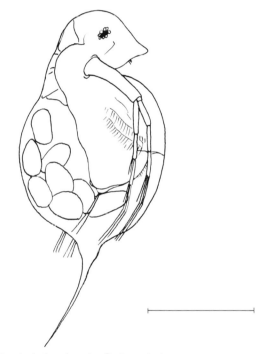

Figure 55 *Daphnia longiremis*, ♀, lateral view

Scale bars = 0.5 mm.

25b. Setae of second antennae do not reach to end of carapace (Figs. 57–59) . . . . . *Daphnia retrocurva*, p. 62

Length 1.2–1.8 mm. High magnification shows enlarged, thickened middle pecten of postabdominal claw. Seta at base of 2nd segment of dorsal ramus of antennae extends beyond end of ramus. Common species.

26a. Metasome and urosome not distinctly separate; 1st antennae shorter than cephalic segment (Fig. 60). . . . . Suborder HARPACTICOIDA, p. 116

Harpacticoid copepods are mainly benthic and littoral organisms and rarely occur in plankton samples. Consult Wilson and Yeatman (1959b) to identify the 9 species occurring in the Great Lakes.

26b. Urosome noticeably narrower than metasome; 1st antennae generally as long or longer than cephalic segment 27

27a. First antennae often reaching to or beyond caudal rami. Body narrows between segment with 5th legs and genital segment (Fig. 2e). . . . . Suborder CALANOIDA, 28

Mature males have geniculate right 1st antennae; adult females often carry a single egg sac.

27b. First antennae usually not reaching past genital segment; body narrows between segments with 4th and 5th legs (5th legs are vestigial) (Figs. 2a, 2b). . . . . Suborder CYCLOPOIDA, 43

Mature males have both 1st antennae geniculate while adult females may carry 2 egg sacs.

28a. Caudal rami elongate, at least 3 times as long as wide 29

28b. Caudal rami not elongate, length less than 3 times width 30

29a. Animals large, up to 3.0 mm; maxillipeds elongate, almost twice body width in lateral view; adult females lack metasomal wings; CVI males have fairly straight 5th legs (Figs. 61, 62) . . . . *Limnocalanus macrurus*, p. 74

Figure 56 Variation in body shape of immature *Daphnia longiremis*

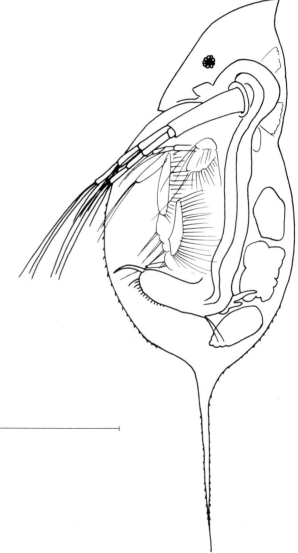

Figure 57 *Daphnia retrocurva*, ♀, lateral view

Scale bars = 0.5 mm.

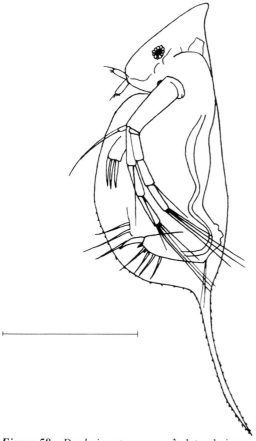

Figure 58 *Daphnia retrocurva*, ♂, lateral view

Figure 59 Variation in helmets of *Daphnia retrocurva*

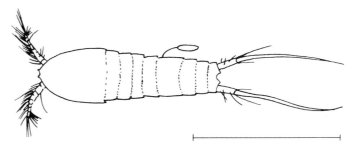

Figure 60 Harpacticoid copepod with spermatophore, ♀, dorsal view

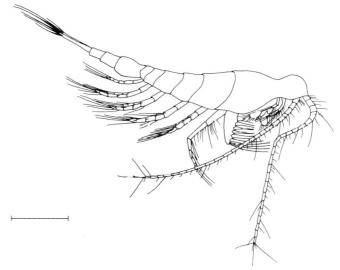

Figure 61 *Limnocalanus macrurus*, ♀, lateral view

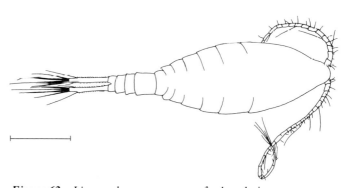

Figure 62 *Limnocalanus macrurus*, ♂, dorsal view

29b. Animals small, up to 1.5 mm; maxillipeds not elongate, subequal to body width in lateral view; adult females have enlarged, pointed metasomal wings; CVI males have bent, hooklike 5th legs . . . . (Figs. 63, 64) . . . . *Eurytemora affinis*, p. 79

30a. Caudal rami with 4 stout terminal setae and a thinner outer seta that is hard to see; large animals, up to 2.9 mm (Figs. 65, 66) . . . . *Senecella calanoides*, p. 73

Unlike most other calanoids, adult females lack 5th legs and do not carry their eggs in egg sacs, and adult males do not have geniculate right 1st antennae.

30b. Caudal rami with 3 or 5 well-developed terminal setae; animal less than 2.5 mm long 31

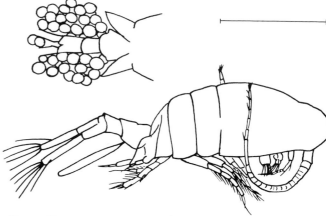

Figure 63 *Eurytemora affinis*, ♀, lateral view

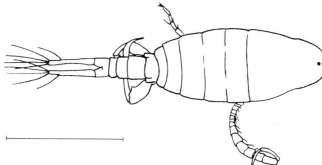

Figure 64 *Eurytemora affinis*, ♂, dorsal view

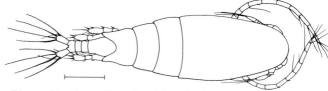

Figure 65 *Senecella calanoides*, ♀, dorsal view

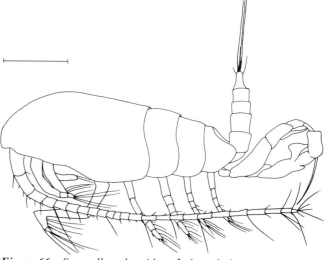

Figure 66 *Senecella calanoides*, ♂, lateral view

Scale bars = 0.5 mm.

Key to Common Great Lakes Crustacean Zooplankton

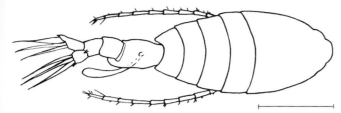

Figure 67 *Epischura lacustris*, ♀, dorsal view

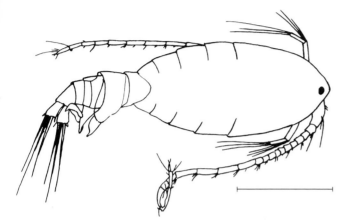

Figure 68 *Epischura lacustris*, ♂, dorsal view

31a. Caudal rami with 3 broad terminal setae (Figs. 67, 68). . . . . *Epischura lacustris*, p. 77

Adult females have twisted urosomes with short, thick spines on the outer corner of the caudal rami. Adult males have smaller spines on the rami and enlarged lateral processes on the right side of the urosome.

31b. Caudal rami with 5 terminal setae and often a slender, dorsally placed inner seta. (Fig. 2e). . . . .
. . . . Family DIAPTOMIDAE, ♂ 38, ♀ 32

Females do not have twisted urosomes, and males lack lateral urosomal processes.

32a. Three urosomal segments 33

32b. Two urosomal segments 37

33a. Genital segment expanded laterally into pointed projections (Fig. 69). . . . .
. . . . ♀ *Leptodiaptomus siciloides*, p. 89

Length 1.0–1.3 mm; 2nd urosomal segment shorter than 3rd; metasomal wings pointed.

33b. Genital segment not expanded laterally; small spines may be present on sides of genital segment 34

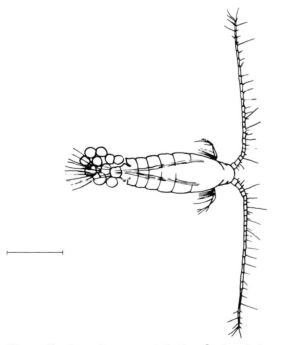

Figure 69 *Leptodiaptomus siciloides*, ♀, dorsal view

Scale bars = 0.5 mm.

34a. Sides of last metasomal segment extended posteriorly into metasomal wings (Figs. 70, 71) 35

34b. Sides of last metasomal segment not extended into wings (Figs. 72, 73) 36

35a. Metasomal wings pointed, triangular (Fig. 70). . . . .
. . . . ♀ *Leptodiaptomus sicilis*, p. 87

Length 1.4–1.9 mm.

35b. Metasomal wings rounded (Fig. 71) . . . .
. . . . ♀ *Skistodiaptomus pallidus*, p. 115

Uncommon in Great Lakes; length 1.0–1.2 mm.

36a. Posterior corners of last segment of metasome rounded (Fig. 72) . . . . ♀ *Skistodiaptomus oregonensis*, p. 91

Length 1.25–1.50 mm; common in Great Lakes.

36b. Posterior corners of last segment of metasome with slight points (Fig. 73) . . . .
. . . . ♀ *Skistodiaptomus reighardi*, p. 115

Length 1.00–1.57 mm; uncommon in Great Lakes

37a. Metasomal wings asymmetrical, left wing longer than right (Fig. 74) . . . .
. . . . ♀ *Leptodiaptomus ashlandi*, p. 82

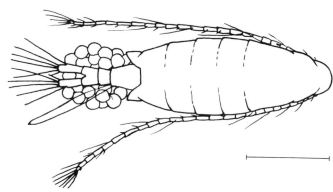

Figure 70   *Leptodiaptomus sicilis*, ♀, dorsal view

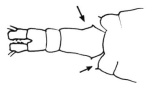

Figure 71   *Skistodiaptomus pallidus*, ♀, urosome, dorsal view

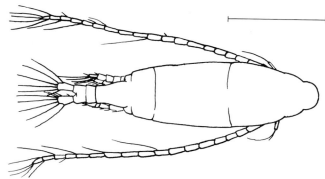

Figure 72   *Skistodiaptomus oregonensis*, ♀, dorsal view

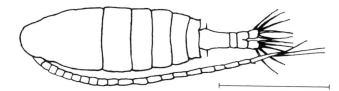

Figure 73   *Skistodiaptomus reighardi*, ♀, dorsal view

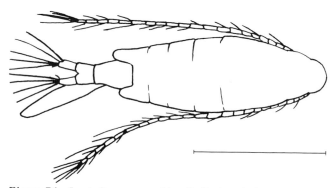

Figure 74   *Leptodiaptomus ashlandi*, ♀, dorsal view

Scale bars = 0.5 mm.

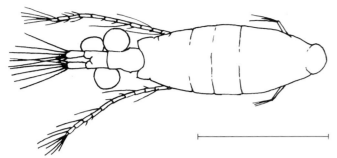

Figure 75 *Leptodiaptomus minutus*, ♀, dorsal view

37b. Metasomal wings symmetrical, rounded, with small spines at corners (Fig. 75) . . . .
. . . . ♀ *Leptodiaptomus minutus*, p. 84

38a. Lateral spine on terminal segment of exopod of right 5th leg subterminal in position (Fig. 76). . . . .
. . . . ♂ *Skistodiaptomus*, pp. 91, 115

High magnification should be used to distinguish the common *S. oregonensis* from the rare *S. pallidus*. The terminal segment of the exopod of the left 5th leg of male *S. pallidus* (fig. 77) has a hooklike process not present in *S. oregonensis* (Fig. 78).

38b. Lateral spine not subterminal in position 39

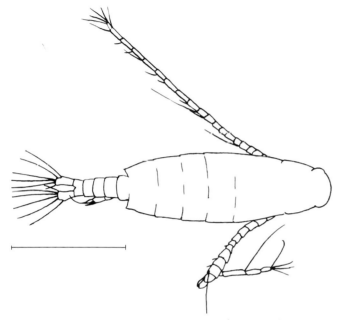

Figure 76 *Skistodiaptomus oregonensis*, ♂, dorsal view

Figure 77 Exopod of left fifth leg of ♂ *S. pallidus*

Figure 78 Exopod of left fifth leg of ♂ *S. oregonensis*

Scale bars = 0.5 mm.

39a. Lateral spine of terminal segment of exopod of right 5th leg large, located on the proximal ⅓ of the segment (Fig. 79). . . ♂ *Leptodiaptomus ashlandi*, p. 82

39b. Lateral spine located between middle and distal ⅓ of the segment; lateral spine large or small 40

40a. Lateral spine of terminal segment of exopod of right 5th leg small, less than ½ width of exopod segment (Fig. 80). . . . . ♂ *Leptodiaptomus minutus* p. 84

40b. Lateral spine large, at least as long as width of exopod 41

41a. Metasomal wings expanded, triangular (Fig. 81). . . . . ♂ *Leptodiaptomus sicilis*, p. 87

41b. Metasomal wings not expanded 42

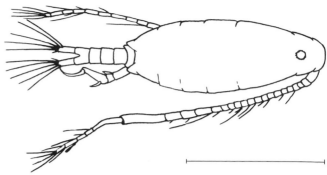

Figure 79   *Leptodiaptomus ashlandi*, ♂, dorsal view

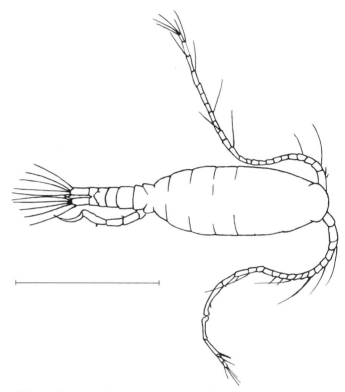

Figure 80   *Leptodiaptomus minutus*, ♂, dorsal view

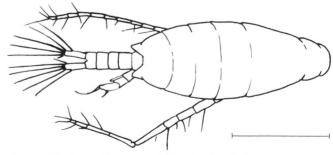

Figure 81   *Leptodiaptomus sicilis*, ♂, dorsal view

Scale bars = 0.5 mm.

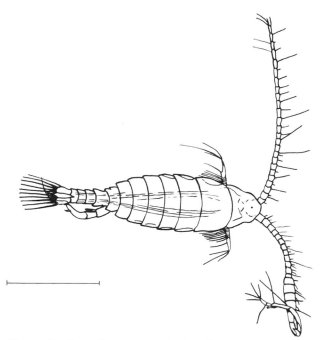

Figure 82 *Leptodiaptomus siciloides*, ♂, dorsal view

Figure 83 Fifth leg of ♂ *Skistodiaptomus reighardi*

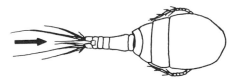

Figure 84 *Macrocyclops albidus*, dorsal view

42a. Terminal claw on exopod of right 5th leg smoothly curved; short process located on antepenultimate segment of right 1st antenna (Fig. 82) . . . . ♂ *Leptodiaptomus siciloides*, p. 82

42b. Terminal claw of exopod of right 5th leg bent, angular. No process on antepenultimate segment of right 1st antenna (Fig. 83). . . . . ♂ *Skistodiaptomus reighardi*, p. 115

43a. Inner seta of caudal ramus long, at least twice the length of the ramus (Fig. 85) 44

43b. Inner seta short, less than twice length of ramus (Fig. 94). 45

44a. Inner seta of caudal ramus less than ½ length of longest caudal seta (Fig. 84) . . . . *Macrocyclops albidus*, p. 115

Length 1–2.5 mm; uncommon in Great Lakes; no hairs on inner margin of rami.

Scale bars = 0.5 mm.

44b. Inner seta longer than ½ length of longest caudal seta. (Figs. 85, 86) . . . . *Mesocyclops edax*, p. 98

Length 0.7–1.5 mm. Common. High magnification shows hairs on inner margin of rami.

45a. First antennae short, less than ¾ length of cephalic segment (uncommon species). 46

45b. First antennae at least ¾ length of cephalic segment 48

46a. First antennae of female with 12 segments (use high magnification). . . . . *Eucyclops prionophorous*, p. 115

Length 0.7–0.9 mm.

46b. First antennae of female with 11 or fewer segments 47

47a. First antennae of female with 8 segments (use high magnification) (Fig. 87) . . . . *Paracyclops fimbriatus poppei*, p. 116

Length 0.7–0.9 mm; uncommon

47b. First antennae of female with 11 segments (use high magnification) (Fig. 88) . . *Diacyclops nanus*, p. 115

Length ♀ 0.45–0.9 mm; uncommon.

48a. Small animals (0.5–0.8 mm); extended 1st antennae reach genital segment (Fig. 89) . . . . *Tropocyclops prasinus mexicanus*, p. 101

48b. Animals longer than 0.7 mm; 1st antennae do not reach genital segment 49

49a. Outer seta of caudal ramus modified into a spine, thicker than inner seta . . . . *Eucyclops*, 50

High magnification shows a row of fine spinules on outer margin of caudal rami of females; 5th leg with only 1 segment.

49b. Outer seta not thicker than inner seta 51

High magnification shows no spinules on outer margin of caudal rami; 5th leg with 2 segments.

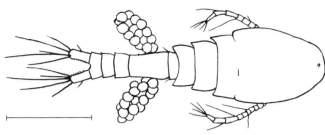

Figure 85 *Mesocyclops edax*, ♀, dorsal view

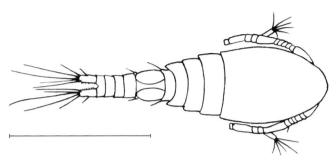

Figure 86 *Mesocyclops edax*, ♂, dorsal view

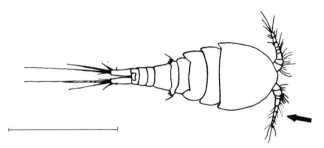

Figure 87 *Paracyclops fimbriatus poppei*, dorsal view

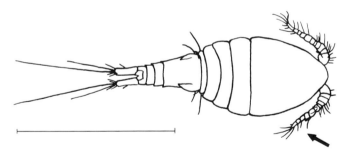

Figure 88 *Diacyclops nanus*, dorsal view

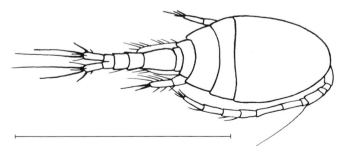

Figure 89 *Tropocyclops prasinus mexicanus*, dorsal view

Scale bars = 0.5 mm.

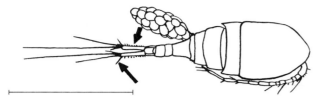

Figure 90 *Eucyclops speratus*, dorsal view

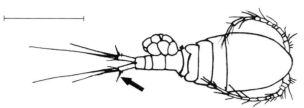

Figure 91 *Eucyclops agilis*, dorsal view

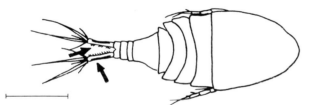

Figure 92 *Cyclops scutifer*, dorsal view

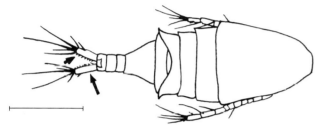

Figure 93 *Cyclops strenuus*, dorsal view

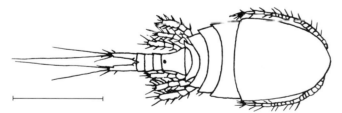

Figure 94 *Acanthocyclops vernalis*, ♀, dorsal view

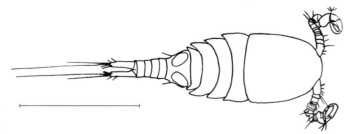

Figure 95 *Acanthocyclops vernalis*, ♂, dorsal view

50a. Caudal rami elongate, length 4–5 times width (Fig. 90) . . . . *Eucyclops speratus*, p. 115

50b. Caudal rami not elongate, length less than 4 times width (Fig. 91) . . . . *Eucyclops agilis*, p. 115

51a. Inner margin of caudal rami with fine hairs (use high magnification; uncommon species) 52

51b. Inner margin of caudal rami without hairs (use high magnification; common species) 53

52a. Caudal rami 4 times as long as wide; 4th and 5th metasomal segments expanded into pointed wings (Fig. 92) . . . . *Cyclops scutifer*, p. 115

Length 0.97–1.29 mm. Smith and Fernando (1978) may also be used to separate this species from *C. strenuus* based on characteristics of the 4th legs.

52b. Caudal rami 5–7 times as long as wide; tips of 4th and 5th metasomal segments with small points, not expanded into wings (Fig. 93) . . . . *Cyclops strenuus*, p. 115

Length 1.2–2.3 mm. See Smith and Fernando (1978) for details on the 4th legs.

53a. Lateral seta located on the posterior ¼ of the caudal ramus (Figs. 94, 95) . . . . *Acanthocyclops vernalis*, p. 93

Length 0.8–1.8 mm. Body more robust than *D. thomasi*. Yeatman (1959) illustrates the 5th legs, which are different than those of *D. thomasi*.

Scale bars = 0.5 mm.

53b. Lateral seta located between the middle and posterior ⅓ of the caudal ramus (Figs. 96, 97) . . . .
. . . . *Diacyclops thomasi*, p. 96

Length 0.9–1.4 mm; body thinner than *A. vernalis*; 5th legs distinct (see Yeatman 1959).

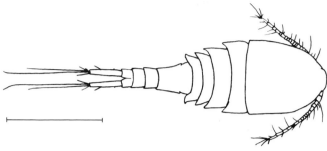

Figure 96 *Diacyclops thomasi*, ♀, dorsal view

Figure 97 *Diacyclops thomasi*, ♂, lateral view

Scale bars = 0.5 mm.

Life History and Ecology of the Major Crustacean Species

CLADOCERANS

Leptodora kindti (Focke 1844)

TAXONOMIC HISTORY

This species was first reported in Germany in 1838 as *Polyphemus kindtii* by Focke and Kindt. It was independently described as *Leptodora hyalina* by Lilljeborg (1860) and as *Hyalasoma dux* by Wagner (1870). The sexual dimorphism of this organism and a limited exchange of information among the investigators resulted in the use of all three names for almost two decades.

In 1867 Muller described both sexes of this cladoceran, leading Sars (1873) to conclude that the three names were synonymous. The taxonomic status of this species was settled in 1880 when Pope renamed it *Leptodora kindtii* Focke. The change in spelling to *Leptodora kindti* is in accordance with the ruling of the 15th International Congress of Zoology (1964).

DESCRIPTION

Unlike most cladocerans, the body of this large, transparent zooplankter is not completely enclosed in a bivalved carapace, and the thoracic legs are not flattened for use in filtering food. *L. kindti* uses its jointed, tubular limbs to grasp prey. Females (Plate 1) possess an enlarged brood pouch that is covered by the reduced carapace. In males (Plate 2) the brood pouch is rudimentary, and the first antennae are elongated.

More detailed descriptions of this organism have been published by Forest (1879), Forbes (1882), Gerschler (1911), and Brooks (1959).

SIZE

The parthenogenetically produced females often grow to over 13 mm while the sexually produced animals are generally less than 9 mm. In Lake Michigan, females produce their first clutch of eggs when they reach a length of 5 to 6.5 mm (Andrews 1949).

In Lake Superior (this study), females average 7.4 mm long (range 6.8–8.7 mm), while males are considerably smaller (average 5.4 mm, range 5.2–6.0 mm).

Leptodora from Lake Michigan have a dry weight of 3.0–19.1 μg (Hawkins and Evans, 1979).

DISTRIBUTION AND ABUNDANCE

L. kindti is found in freshwater lakes, ponds, and occasionally rivers in the northern hemisphere. It is generally re-

TABLE 1
Reports of *Leptodora kindti* in the Great Lakes

	Sampling Date	Abd[a]	Reference		Sampling Date	Abd[a]	Reference
LAKE ERIE	1918	P	Langlois 1954	**LAKE HURON**	1903, 1905, 1907	P	Bigelow 1922
	1918–1920	F	Clemens and Bigelow 1922		1907	C	Sars 1915
	1928–1930	P	Wright 1955		1967–1968	P	Patalas 1972
	1928	F	Sibley 1929		1970–1971	P	Carter 1972
	1928–1929	P	Wilson 1960		1974	P	Basch et al. 1980
	1929	F	Ewers 1933		1974–1975	P	McNaught et al. 1980
	1929	F	Kinney 1950				
	1938–1939	P	Chandler 1940	**LAKE ONTARIO**	1919, 1920	U	Clemens and Bigelow 1922
	1946–1948	P	Andrews 1949		1939–1940	U	Tressler et al. 1953
	1950–1951	P	Davis 1954		1967	P	Patalas 1969
	1956–1957	C	Davis 1962		1967–1968	P	Patalas 1972
	1961	P	Britt et al. 1973		1969–1970, 1975	P	McNaught and Hasler 1966
	1962	CF	Wolfert 1965		1970	C	Watson and Carpenter 1974
	1967–1968	P	Davis 1968		1972	U	Czaika 1974a
	1967	P	Patalas 1972		1972–1973	U	Czaika 1978a
	1970	A	Watson and Carpenter 1974				
	1971–1972	P	Rolan et al. 1973	**LAKE SUPERIOR**	1871	F	Smith 1874b
LAKE MICHIGAN	1881	P	Forbes 1882		1889	P	Forbes 1891
	1887–1888	U	Eddy 1927		1893	P	Birge 1893
	1926–1927	R	Eddy 1934		1964	P	Olson and Odlaug 1966
	1954–1955, 1958	—	Wells 1960		1967–1969	P	Swain et al. 1970b
	1954–1961	UF	Wells and Beeton 1963		1967–1968	P	Patalas 1972
	1966, 1968	P	Wells 1970		1971–1972	P	Selgeby 1975a
	1969–1970	P	Gannon 1972a		1971	P	Selgeby 1974
	1969–1970	P	Gannon 1974		1973	P	Watson and Wilson 1978
	1969–1970	U	Gannon 1975		1973	P	Upper Lakes Ref. Group 1977
	1971	C	Howmiller and Beeton 1971		1974	P	Basch et al. 1980
	1971–1972	P	Beeton and Barker 1974		1979–1980	P	This study
	1971–1977	P	Evans et al. 1980				
	1973	P–C	Stewart 1974				
	1973–1974	U	Torke 1975				
	1975–1977	—	Hawkins and Evans 1979				

[a] Abundance Code
R = rare   U = uncommon   P = present   C = common   A = abundant   F = found in fish stomach contents
— = abundance ranking not appropriate

stricted to water bodies located between 35 degrees and 60 degrees north latitude.

L. kindti is commonly found in all of the Great Lakes (Table 1) but in relatively low numbers. Abundance estimates are variable due to this animal's patchy horizontal distribution (Chandler 1940) and its ability to avoid plankton traps and small nets (Tressler et al. 1953). Abundance is generally less than 100 m[-3] (McNaught and Buzzard 1973; Howmiller and Beeton 1971; present study). In Lake Superior the average annual density is less than 1 m[-3] (Selgeby 1975a), but swarms of 200 m[-3] (Stewart 1974) and 2449 m[-3] (Rolan et al. 1973) have been reported from lakes Michigan and Erie, respectively.

LIFE HISTORY IN THE GREAT LAKES

In the Great Lakes *L. kindti* is commonly seen from May through November or December (Chandler 1940; Andrews 1949; Rolan et al. 1973; Stewart 1974; Selgeby 1974). Abundance peaks in midsummer or early fall (Andrews

1949; Davis 1954; Wells 1960; Beeton and Barker 1974; Gannon 1974; Watson and Carpenter 1974).

In the fall males and females mate and a resting egg is formed, which overwinters. In the spring, generally in April or May, a 1-mm nauplius hatches from each egg (Warren 1901). The nauplii grow to 1.5–2.0 mm and then molt to an immature female form. These sexually produced females possess a secondary eyespot located behind and below the compound eye. As the animals mature the eyespot fades. This first generation is not very numerous, and adults are smaller and produce fewer clutches with fewer eggs than the succeeding generations.

The sexually produced females reproduce parthenogenetically and deposit 0.38- to 0.40-mm diameter eggs into the brood sac on their back. The eggs develop directly into immature females (no naupliar stages are present in parthenogenetically produced animals), which are released from the brood chamber when the parent molts. The first parthenogenetically produced females usually mature in June, and reproduce parthenogentically throughout the summer.

In the fall, some of the females produce eggs that hatch into males. The stimulus for male production is not known, but it may be linked to blue-green algae abundance (Andrews 1948). The males mate with females to produce the resting eggs.

Females usually outnumber males, except during October and November when sexual reproduction is taking place.

Andrews (1949) constructed a size-frequency distribution curve of 1300 parthenogenetically produced females collected over a two-year period. The curve displayed extreme irregularity and lacked distinct modes, thus failing to indicate the number of adult instars produced in a natural population.

ECOLOGY IN THE GREAT LAKES

Habitat. L. kindti occurs in both the littoral and limnetic zones of most lakes. It has an uneven horizontal distribution and has been observed forming large patches or swarms (Andrews 1948; Wilson 1960; Swain et al. 1970a; Beeton and Barker 1974; Gannon 1975).

This species may be temperature limited (Andrews 1948). Its abundance declines steadily in October and November as the water cools, even though the food supply remains at a relatively high level for several more weeks.

Diurnal Migrations. L. kindti migrates upward as light intensity decreases, arriving at the surface at or just after sunset. The animals remain near the surface until early morning (Andrews 1949; Wells 1960). Immature forms precede the adults to the surface at night and remain there later in the morning.

Andrews (1949) was unable to collect any organisms during the day, possibly because they were able to see and avoid his net. Further work by Wells (1960) showed that many Leptodora are located near the lake bottom during the day, although some animals tend to remain in the metalimnion. The depth of daytime distribution seems dependent on the depth of the water column.

Feeding Ecology. Although L. kindti is a large zooplankter with legs modified for raptorial feeding, it is a feeble swimmer and waits for prey to happen by instead of actively pursuing it. Andrews (1949) conducted laboratory feeding trials and discovered that this species captures the slowest-moving prey items first when offered a choice of prey. To be utilized as a food source, fast-moving prey must be very numerous. The most common zooplankters eaten include cyclopoids, diaptomids, Epischura, daphnids, and Eurycercus.

As a group, cladocerans may be the major food item of L. kindti (Mordukhai-Boltovskoi 1958). Stewart (1974) found that the peak abundance of L. kindti coincided with the dominance of Daphnia and high abundance of other larger cladocerans. Andrews (1949) showed that the abundance of L. kindti decreased concurrently with a decrease in abundance of other cladocerans and an increase in phytoplankton density.

McNaught et al. (1980) found that Leptodora filters nannoplankton in Lake Huron at the slow rate of 0.004–0.006 ml · animal^{-1} · hour^{-1}.

As Food for Fish. Its large size and tendency to swarm make L. kindti an ideal food for fish. It is a major item in the diet of bass, trout-perch, shiners, crappie, sauger, walleye, perch, cisco, and freshwater drum (Clemens and Bigelow 1922; Sibley 1929; Ewers 1933; Kinney 1950; Wolfert 1965; Engel 1976). Without its extreme transparency and pattern of vertical migration, Leptodora would probably be eliminated from most lakes.

LIFE HISTORY AND ECOLOGY IN OTHER LAKES

L. kindti is common in many large lakes (Carl 1940) and generally has a life history similar to that described previously (Marsh 1897; Birge 1897). The occurrence of large swarms and patchy distribution patterns reported from Green Lake in Wisconsin (Marsh 1897) are similar to those observed in the Great Lakes. However, there are some exceptions to the generalized pattern of vertical migration described previously. Forbes (1891) occasionally found *L. kindti* at the surface during the day, while in clear, shallow lakes the animals were sometimes found in the mud.

These organisms can be difficult to culture in the laboratory (Engel 1976). It is important to provide the animals with an abundant food supply and to prevent injuries due to contact with the aquarium sides. Temperature, light, pH, aeration, and circulation of culture media are not primary factors responsible for laboratory die-offs (Andrews 1949).

Young *L. kindti* are voracious feeders immediately after release from the brood pouch. In feeding studies, Birge (1897) observed that *L. kindti* seemed to squeeze out and swallow the interior of the prey items, particularly *Cyclops* and *Daphnia*. Cannibalism may occur in some lab cultures when food is limiting (Andrews 1949).

Polyphemus pediculus (Linné 1761)

TAXONOMIC HISTORY

Polyphemus pediculus is the sole freshwater representative of the family Polyphemidae in North America. This family is represented by several other genera and species in the Caspian Sea region.

DESCRIPTION

Ischregt (1933), Scourfield and Harding (1941), Brooks (1959), and Buttorina (1968) provide detailed descriptions of this unusual cladoceran. *Polyphemus* is easily recognized by its reduced carapace, large eye, globular brood chamber, and elongate abdomen terminating in two long caudal setae (Plate 3). Males are distinguished from females by a hook on the inner surface of the first leg (Fig. 98) (Mordukhai-

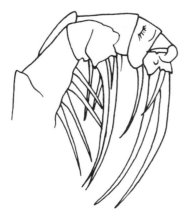

Figure 98   First leg of *Polyphemus pediculus*

Boltovskoi 1965), although this feature is difficult to see even under high magnification.

SIZE

Females may reach a length of 1.5–2 mm while males are smaller, generally reaching only 0.8 mm (Scourfield and Harding 1941; Brooks 1959). In Lake Superior, we found female *Polyphemus* to be fairly small (0.74–1.08 mm), averaging 0.89 mm. This cladoceran has a dry weight of 2.2–4.1 μg (Hawkins and Evans 1979).

DISTRIBUTION AND ABUNDANCE

P. pediculus is characteristic of northern or Arctic water bodies (Gurney 1923), occurring in pools, marshes, and along lake margins throughout the northern hemisphere. It has been collected from all of the Great Lakes (Table 2). It generally occurs in low densities (< 20 m^{-3}) although abundance may increase for a brief period in July or August to 170–1000 organisms m^{-3} (Olson and Odlaug 1966; Wells 1970; Stewart 1974). Recent changes in relative abundances led Gannon (1970) to speculate that *P. pediculus* is replacing *Leptodora kindti* (another predatory cladoceran) in Lake Michigan.

LIFE HISTORY IN THE GREAT LAKES

Most Great Lakes investigators have only observed *P. pediculus* in their mid-to-late summer samples (Wells 1960; Patalas 1969; Carter 1972; Gannon 1972a, 1974, 1975; Selgeby 1974; Czaika 1974a; Stewart 1974; Torke 1975).

Polyphemus produces two generations each year. The summer population of females in Lake Superior hatches from resting eggs that overwintered (Selgeby 1974). These females mature and produce parthenogenetic eggs that develop into sexual males and females. The ephippial or resting eggs are produced by the matings of this generation.

ECOLOGY IN THE GREAT LAKES

Habitat.   This cladoceran is generally most abundant in inshore areas (Stewart 1974), where it may exhibit a patchy distribution pattern (Swain et al. 1970b). *P. pediculus* shows a preference for the upper water strata a few meters below the surface but has been found to depths of 40 m (Wells 1960; Olson and Odlaug 1966).

Diurnal Migration.   Migration studies of this species are limited by its sparse abundance in most areas. Wells (1960) found evidence of migration towards the surface at sunset or slightly before. A slight evening descent may precede a sec-

TABLE 2
Reports of *Polyphemus pediculus* in the Great Lakes

	Sampling Date	Abd[a]	Reference		Sampling Date	Abd[a]	Reference
LAKE ERIE	—	—	Reighard 1894	**LAKE HURON (continued)**	1970–1971	P	Carter 1972
	—	—	Price 1963		1974	P	Basch et al. 1980
					1974–1975	P	McNaught et al. 1980
LAKE MICHIGAN	1954–1955, 1958	P	Wells 1960	**LAKE ONTARIO**	1967	U	Patalas 1969
	1954–1961	U	Wells and Beeton 1963		1967–1968	P	Patalas 1972
	1966, 1968	P	Wells 1970		1972	R	Czaika 1974a
	1969–1970	U	Gannon 1972a				
	1969–1970	U	Gannon 1974	**LAKE SUPERIOR**	1889	P	Forbes 1891
	1969–1970	U	Gannon 1975		1893	P	Birge 1893
	1971–1972	P	Beeton and Barker 1974		1964	P	Olson and Odlaug 1966
	1972–1977	P	Evans et al. 1980		1967–1968	U	Patalas 1972
	1973	P–C	Stewart 1974		1969	P	Swain et al. 1970b
	1973–1974	U	Torke 1975		1971	P	Selgeby 1974
	1974	P	Evans and Stewart 1977		1971–1972	U	Selgeby 1975a
	1975–1979	—	Hawkins and Evans 1979		1973	P	Watson and Wilson 1978
					1973	P	Upper Lakes Ret. Group 1977
LAKE HURON	1903, 1905, 1907	C	Bigelow 1922		1974	P	Basch et al. 1980
	1907	C	Sars 1915		1979–1980	P	This study
	1967–1968	U	Patalas 1972				

[a]Abundance Code
R = rare   U = uncommon   P = present   C = common   A = abundant   F = found in fish stomach contents
— = abundance ranking not appropriate

ondary peak near the surface just before dawn. The animals then descend to their daytime depths of a few meters below the surface.

Feeding Ecology. *P. pediculus* is a predatory cladoceran. Its modified thoracic appendages enable it to grasp small prey items including protozoans, rotifers, and other cladocerans (Brooks 1959; Pennak 1978). Laboratory studies (Anderson 1970) show no evidence that it consumes nauplii of diaptomid or cyclopoid copepods. The food items consumed are broken down by enzymes capable of digesting fats, carbohydrates, and proteins (Hasler 1937).

The distribution of *P. pediculus* may be affected by its food habits. In Lake Michigan, it is found concentrated in inshore areas where the zooplankton populations are dominated by small prey organisms such as rotifers and *Bosmina longirostris* (Stewart 1974).

It was also found to consume nannoplankton in Lake Huron (McNaught et al. 1980). A yearly average of algae equivalent to 22.7% of its body weight is consumed each day, with filtering rates varying from 0.0015–0.044 ml · animal^{-1} · hour^{-1}.

As Food for Fish and Other Organisms. *P. pediculus* does not appear to be a common prey item for most fish. It has only been found in small quantities in the stomachs of young bloater (Wells and Beeton 1963) and arctic char (Langeland 1978). Although cyclopoid copepods consume other cladocerans, there is no indication that they attack *P. pediculus* (Anderson 1970).

LIFE HISTORY AND ECOLOGY IN OTHER LAKES

The distribution and ecology of this species have been studied in lakes in British Columbia (Carl 1940) and southern Saskatchewan (Moore 1952).

Laboratory experiments (Hutchinson 1967) show that *P. pediculus* is negatively phototactic in a strong horizontal light beam but becomes positively phototactic when the intensity of the beam is decreased. The nutritional needs and development of its embryos in the brood chamber have also been investigated (Patt 1947).

Diaphanosoma (Fischer 1850): *D. birgei* and *D. brachyurum*

TAXONOMIC HISTORY

The animals in this group were originally described as members of the genus *Diaphanosoma* but were changed to the genera *Sida* and *Daphnella* before *Diaphanosoma* was reestablished as a true genus (Herrick 1884; Richard 1895). Two of the 11 species in this genus, *D. leuchtenbergianum* (Fischer 1850) and *D. brachyurum* (Liéven 1848) are reported from the Great Lakes. These two forms have been considered ecotypic varieties of a single species (Brooks 1959), with *D. leuchtenbergianum* being the more limnetic variety.

Recent work by Kořínek (1981) shows that, based on its original description, the name *D. leuchtenbergianum* is synonymous with *D. brachyurum*, but there are two species of *Diaphanosoma* in North America. He proposes the name *D. birgei* for American specimens of the more limnetic species and retains the name *D. brachyurum* for the more littoral species. We follow this name change.

DESCRIPTION

This genus is distinguished from other members of the family Sididae by characteristics of the second antennae (Brooks 1959). The two Great Lakes species are recognized by their enlarged second antennae, enlongate bodies that lack terminal shell spines, and rounded heads without crests (Plates 4, 5). The more limnetic species, now called *D. birgei* has previously been distinguished from *D. brachyurum* by its larger head (head equals ⅔ length and ½ width of carapace), smaller eye, which is located near the midventral margin of the head, and more transparent body color (Figs. 99, 100). The second antennae of *D. birgei*, which reach the posterior margin of the carapace when reflexed, are longer than those of *D. brachyurum* (Fordyce 1900; Brooks 1959; Brandlova et al. 1972; Pennak 1978). These features are somewhat variable, and Kořínek (1981) proposes using the spine on the distal margin of the first segment on the exopod (longer branch) of the second antennae to differentiate the species. The spine extends past the tip of the segment in *D. birgei* but not in *D. brachyurum*.

Males of both species are distinguished from females by their elongate antennules and hooked first legs.

SIZE

Brooks (1959) reported that *D. brachyurum* was slightly smaller than *D. birgei*:

Figure 99 Head shape of *Diaphanosoma birgei*

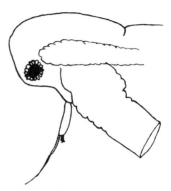

Figure 100 Head shape of *Diaphanosoma brachyurum*

D. brachyurum	♀ 0.8–0.9 mm	♂ 0.4 mm
D. birgei	♀ 0.9–1.2 mm	♂ 0.8 mm

The female *D. birgei* we collected from Lake Superior averaged 1.02 mm long (range 0.92–1.12 mm).

Hawkins and Evans (1979) found that *Diaphanosoma* from lake Michigan have a dry weight of 1.2–3.6μg.

DISTRIBUTION AND ABUNDANCE

D. birgei has been collected in North America from the Arctic Circle south to Louisiana. It is common in northern Europe and Asia (Scheffer and Robinson 1939; Scourfield and Harding 1941; Pennak 1949; Moore 1952; Comita 1972; Binford 1978; and Kořínek 1981). *D. brachyurum* has a similar distribution pattern but is also found in Africa and Central and South America (Birge 1897; Gurney 1923; Brandlova et al. 1972; Lewkowicz 1974).

D. birgei is more widely distributed in the Great Lakes than the littoral *D. brachyurum* (Table 3). The former species is generally found in low numbers (< 1 m⁻³) in Lakes Ontario, Michigan, and Superior (Patalas 1969; Czaika 1974a; Gannon 1974; Stewart 1974; Selgeby 1975a) although peaks of 300 m⁻³ were found in Lake Superior's Chequamegon Bay (Selgeby 1974). It is fairly common in Lake Huron's Georgian Bay (Carter 1972) and in Lake Erie, where average summertime densities of 400 m⁻³ and peaks of 2400 m⁻³ have been recorded (Davis 1954).

TABLE 3
Reports of *Diaphanosoma* in the Great Lakes

	Sampling Date	Abd[a]	Reference		Sampling Date	Abd[a]	Reference
LAKE ERIE	1919–1920	F	Wickliff 1920	**LAKE MICHIGAN (continued)**	1971–1972	1U	Beeton and Barker 1974
	1919–1920	2F	Clemens and Bigelow 1922		1971–1973	1U	Stewart 1974
	1928–1929	1P–C, 2P	Wilson 1960		1972–1977	1P	Evans et al. 1980
	1928–1930	1P	Wright 1955		1973–1974	1P	Torke 1975
	1929	1F	Ewers 1933		1975–1977	—	Hawkins and Evans 1979
	1938–1939	1P	Chandler 1940				
	1948–1949, 1959	1C	Bradshaw 1964	**LAKE HURON**	1903, 1905, 1907	1U, 2U	Bigelow 1922
	1950–1951	1C	Davis 1954		1968	1U	Patalas 1972
	1956–1957	1P	Davis 1962		1970–1971	1C	Carter 1972
	1961	1P	Britt et al. 1973		1970	1R	Watson and Carpenter 1974
	1967	1P	Davis 1968		1974–1975	1U	McNaught et al. 1980
	1967–1968	1P	Davis 1969				
	1968, 1970	1P	Heberger and Reynolds 1977	**LAKE ONTARIO**	1967	1U	Patalas 1969
	1968	1P	Patalas 1972		1967	1P	Patalas 1972
	1970	1C	Watson and Carpenter 1974		1972	1R	Czaika 1974a
	1971–1972	1C	Rolan et al. 1973		1972–1973	1R	Czaika 1978a
LAKE MICHIGAN	1888	1C	Eddy 1927	**LAKE SUPERIOR**	1928	2R	Eddy 1934
	1954–1955, 1958	2P	Wells 1960		1968	1U	Patalas 1972
	1954, 1966, 1968	2R	Wells 1970		1971	1P	Selgeby 1974
	1969–1970	1P	Gannon 1972a		1971–1972	1U	Selgeby 1975a
	1969–1970	1P, 2P	Gannon 1974		1973	1U	Upper Lakes Ref. Group 1977
	1969–1970	1P	Gannon 1975		1979–1980	1P	This study

[a] Abundance Code

1 = *D. birgei*   2 = *D. brachyurum*   R = rare   U = uncommon   P = present   C = common   A = abundant

F = found in fish stomach contents   — = abundance ranking not appropriate

LIFE HISTORY IN THE GREAT LAKES

In Lake Erie *D. birgei* is commonly collected between May and October (Chandler 1940; Wright 1955; Davis 1962; Britt et al. 1973; Watson and Carpenter 1974). In the other Great Lakes this species is generally found only in the fall (Beeton and Barker 1974; Czaika 1974a; Watson and Carpenter 1974; Gannon 1975). Abundance peaks have usually been reported during the fall months (Eddy 1927; Davis 1954 and 1962; Bradshaw 1964; Patalas 1969; Rolan et al. 1973; Britt et al. 1973; Stewart 1974).

The life history of this genus in the Great Lakes is not well documented. Summer reproduction appears to be parthenogenetic. Small numbers of males appear only in the fall. In Lake Michigan (Stewart 1974) males have been collected during the September population peak.

ECOLOGY IN THE GREAT LAKES

Habitat. *D. brachyurum* is found in littoral areas while *D. birgei* is more limnetic (Brooks 1959). In the Great Lakes *D. birgei* is usually found in the open waters of sheltered bays, harbors, and island areas rather than in the deep offshore regions (Patalas 1969; Gannon 1972a; Carter 1972; and Selgeby 1974). In Ohio lakes this species was most common at a depth of 3–6 m and was seldom collected at the surface (Winner and Haney 1967).

In Lake Erie *Diaphanosoma* is most abundant in the western basin in July and in the eastern and central basins during the fall (Davis 1968, 1969). It is not known if eutrophication and high temperatures are responsible for this seasonal change in distribution.

Diurnal Migrations. Wells (1960) reported that *D. brachyurum* was found at all depths in Lake Michigan but preferred the upper layers, with the animals more concentrated at the surface at night than during the day.

Feeding Ecology. Although the feeding behavior of this genus has not been studied in the Great Lakes, Gliwicz (1969) found that other populations of *D. brachyurum* are filter feeders capable of ingesting particles from 5–154μm. The smallest particles (1–5μm) are consumed in the greatest quantities while very few particles greater than 18μm are ingested. Stomach analysis showed chlorophytes (7–15μm) and diatoms (12μm) are the principal items consumed.

As Food for Fish. In the Great Lakes *Diaphanosoma* are consumed by largemouth bass, cisco, crappie, logperch, shiners, yellow perch, and other fish (Wickliff 1920; Clemens and Bigelow 1922; Ewers 1933; Wilson 1960).

LIFE HISTORY AND ECOLOGY IN OTHER LAKES

D. birgei has been studied in Lake Washington (Scheffer and Robinson 1939), British Columbia (Carl 1940), Colorado (Pennak 1949), the midwestern United States (Birge 1897; Marsh 1897; Comita 1972), and in the Atchafalaya River in Louisiana (Binford 1978). Most studies reported seasonal abundance patterns similar to those of the Great Lakes.

Diaphanosoma species have been found to prefer epilimnetic waters (Marsh 1897; Engel 1976) and migrate to the surface at night in several lakes (Marsh 1897; Brandlova et al. 1972).

Field studies employing grazing chambers (Lane 1978) showed that *D. birgei* is readily preyed upon by omnivorous zooplankton including *Diacyclops thomasi*, *Chaoborus*, *Mesocyclops edax*, *Leptodora kindti*, *Tropocyclops prasinus mexicanus*, *Skistodiaptomus oregonensis*, and *Leptodiaptomus minutus*. In a separate study, Anderson (1970) found that *Diaphanosoma leuchtenbergianum* (*D. birgei*?) was the most common prey item of *Diacyclops thomasi* in Patricia Lake, Alberta.

Holopedium gibberum (Zaddach 1855)

TAXONOMIC HISTORY

Holopedium gibberum was first described in 1855 by Zaddach. It is one of the two species of the genus *Holopedium*

of the family Holopedidae. The distinct characteristics of this family, genus, and species have resulted in very few changes in the taxonomy of this organism.

DESCRIPTION

H. gibberum is easily recognized by the large gelatinous mantle covering its humped carapace. This mantle is often twice as long as the body, thus increasing the animal's volume by eight times. In live plankton samples, the mantle generally appears as a small glob of jelly; in preserved samples it is often lost. *H. gibberum* swims upside down with its second antennae and filtering appendages projecting through a T-shaped opening in the ventral surface of the mantle. The second antennae are uniramous in females (Plate 6) and biramous in males.

SIZE

Brooks (1959) reports females from 1.5–2.2 mm (excluding mantle) and males from 0.5–0.6 mm. In Lake Superior we found the females to be a bit smaller (1.0–1.7 mm). Without the mantle, *Holopedium* from Lake Michigan have a dry weight of 1.9–10.9μg (Hawkins and Evans 1979).

DISTRIBUTION AND ABUNDANCE

H. gibberum occurs in northern and arctic lakes of Europe and North America. In the United States it is restricted to the northern states and the mountainous areas from California to Colorado (Gurney 1923; and Brooks 1959). Carpenter (1931) regarded this species as a glacial relict. It has been found in brackish water, but it is less abundant and shows limited reproduction there.

H. gibberum is found in low numbers in all the Great Lakes (Table 4). Davis (1968) and Patalas (1972) reported this species from the Eastern and Central basins of Lake Erie but not from the Western Basin. It exhibits a very patchy distribution, generally numbering less than 6 m⁻³ but occasionally concentrating in numbers over 4000 m⁻³ (Wells 1970; Stewart 1974; McNaught et al. 1980). Abundance is generally high in Georgian Bay of Lake Huron (Sars 1915; Bigelow 1922; Carter 1972; Watson and Carpenter 1974) but is lower along the western side of Lake Huron (7–17 m⁻³), (Basch et al. 1980) and in Lake Superior (35 m⁻³) (Selgeby 1975a).

LIFE HISTORY IN THE GREAT LAKES

This cladoceran is usually absent from winter and spring plankton samples (Wells 1960; Davis 1969; Beeton and Bar-

TABLE 4
Reports of *Holopedium gibberum* **in the Great Lakes**

	Sampling Date	Abd[a]	Reference		Sampling Date	Abd[a]	Reference
LAKE ERIE	1919–1920	F	Clemens and Bigelow 1922	**LAKE HURON (continued)**	1967–1968	P	Patalas 1972
	1928–1929	U	Wilson 1960		1970–1971	A	Carter 1972
	1967	C	Davis 1968		1970	P	Watson and Carpenter 1974
	1967–1968	P	Davis 1969		1974	P	Basch et al. 1980
	1967–1968	P	Patalas 1972		1974–1975	P	McNaught et al. 1980
	1970	P	Watson and Carpenter 1974				
				LAKE ONTARIO	1967	P	Patalas 1969
LAKE MICHIGAN	1881	P	Forbes 1882		1967–1968	U	Patalas 1972
	1954–1955, 1958	C	Wells 1960		1970	P	Watson and Carpenter 1974
	1954–1961	PF	Wells and Beeton 1963		1972	U	Czaika 1974a
	1966, 1968	P	Wells 1970		1973	R	Czaika 1978a
	1969–1970	P	Gannon 1975				
	1969–1970	C	Gannon 1972a	**LAKE SUPERIOR**	1889	P	Forbes 1891
	1969–1970	P	Gannon 1974		1893	P	Birge 1893
	1971–1972	A	Beeton and Barker 1974		1967–1968	U	Patalas 1972
	1972–1977	P	Evans et al. 1980		1970–1971	P	Conway et al. 1973
	1973	P	Stewart 1974		1971–1972	P	Selgeby 1975a
	1973–1974	P	Torke 1975		1971	P	Selgeby 1974
	1975–1977	—	Hawkins and Evans 1979		1973	P	Watson and Wilson 1978
					1973	P	Upper Lakes Ref. Group 1977
LAKE HURON	1903, 1905, 1907	A	Bigelow 1922		1974	C	Basch et al. 1980
	1907	C	Sars 1915		1979–1980	P	This study

[a] Abundance Code
R = rare   U = uncommon   P = present   C = common   A = abundant   F = found in fish stomach contents
— = abundance ranking not appropriate

ker 1974). Abundance increases during the summer, with population peaks occurring between June and October (Wells 1960; Patalas 1969; Gannon 1972a; Stewart 1974; Torke 1975; Selgeby 1975a).

The reproductive pattern of this species is not well documented for the Great Lakes. Selgeby (1974, 1975a) found very few ovigerous females in Lake Superior even though the population increased from July to August. He observed one male and a few ephippial females in October. In Lake Michigan, males have been found to comprise 25% of the fall population (Torke 1975). These data suggest that female *H. gibberum* hatch from resting eggs and produce a sexual generation. Those animals mate and produce ephippial eggs that overwinter.

ECOLOGY IN THE GREAT LAKES

Habitat. *H. gibberum* has been found in both the littoral (Forbes 1891) and pelagic (Birge 1893) regions of lakes. Carter (1972) found greatest numbers inshore in June with the population peaking offshore during the summer. This may be due to its preference for cold water. It is considered a cold water stenotherm (Pennak 1978) and occurs primarily in oligotrophic soft-water lakes (Thienemann 1926, Pejler 1965). However, it is often abundant in the relatively hard waters of Lakes Huron and Michigan.

Diurnal Migration. The vertical distribution of *Holopedium* has been studied in Lake Michigan (Wells 1960). This cladoceran remains in the upper 20 m of the water column above the metalimnion during the day, moving toward the surface beginning in late afternoon. Greatest abundance at the surface was observed two hours before sunset in July and at sunset in October. The animals descend slightly for the night.

Feeding Ecology. *H. gibberum* is an omnivorous filter feeder, consuming several species of planktonic algae as it swims about upside down (Stenson 1973). Although considered a generalist (Allan 1973), *Holopedium* in Lake

Huron was found to consume only nannoplankton approximately 22μm long during August and October (McNaught 1978). This animal is a very effective grazer at low densities of phytoplankton and may be adapted for feeding in areas of dispersed phytoplankton (McNaught 1978). In Lake Huron, *Holopedium* filters nannoplankton at a rate of $0.129-0.147$ ml $\cdot$ animal^{-1} $\cdot$ hour^{-1} (McNaught et al. 1980) and ingests an average of 0.8% of its body weight each day.

As Food for Fish. *H. gibberum* is occasionally eaten by cisco (Clemens and Bigelow 1922; Wilson 1960), bloater (Wells and Beeton 1963), the Arctic char (Langeland 1978), and coho salmon (Engel 1976). The small actual body size of this animal helps it avoid most vertebrate predation, while its large invisible gelatinous mantle offers protection from invertebrate predation (Allan 1973). McNaught (1978) reported that *H. gibberum* may taste bad to fish, possibly due to the chemical composition of the mantle.

LIFE HISTORY AND ECOLOGY IN OTHER LAKES

The mantle, which covers most of the body of *Holopedium*, is composed of one or more acid muco-polysaccharides (Hamilton 1958). It is believed to be secreted by the outer epithelium of the carapace and may have other functions besides protecting the animal from predation. Hamilton (1958) suggested that the mantle may reduce the animal's density and help retard its rate of sinking.

In most areas female *Holopedium* hatch from resting eggs in the spring when the water temperature reaches $4-7$ degrees C (Hamilton 1958; Tash 1971; and Stenson 1973). The animals mature in approximately 6 weeks and reproduce parthenogenetically in June and July. Each female can carry up to 20 eggs that develop into males and sexual females by August. These animals mate and produce the ephippial eggs. In most areas this species disappears between September and November. In warmer lakes *Holopedium* may produce several parthenogenetic generations before sexual animals are produced (Tash 1971).

H. gibberum is usually found in slightly acidic waters with a pH of $6.0-6.8$ (Carpenter 1931) although it can withstand a pH of $4.0-7.5$ (Hamilton 1958). It is restricted to soft-water lakes with less than 20 mg/1 Ca^{++} and temperatures less than 25 degrees C (Hamilton 1958). This species is often found in large swarms (Hamilton 1958; Stenson 1973) swimming about in the epilimnion (Engel 1976; Langford 1938).

In most lakes *Holopedium* shows a regular pattern of diurnal migration. However in one lake in Colorado, Pennak (1944) found the maximum abundance occurring at the surface at noon.

Ceriodaphnia spp. (Dana 1853)

TAXONOMIC HISTORY

Five species of *Ceriodaphnia* have been collected from the Great Lakes: *C. lacustris* Birge 1893, *C. laticaudata* P. E. Müller 1867, *C. pulchella* Sars 1862, *C. quadrangula* (O. F. Müller) 1785 and *C. reticulata* (Jurine) 1820. A sixth species, *Ceriodaphnia scitula*, was reported in Lake Huron by Sars (1915). However, the description Sars gave of this new species conformed to that of *C. lacustris*.

DESCRIPTION

All species of *Ceriodaphnia* possess spherical carapaces without an elongate posterior spine (Plate 7). The small but distinct head is bent ventrally and contains a large compound eye. The antennules are reduced in size. Males are distinguished from females by their modified first legs and long, stout setae on the antennules.

Brooks (1959) and Brandlova et al. (1972) should be consulted when identifying *Ceriodaphnia* to species. High magnification may be necessary to determine the characteristics of the postabdominal claw and anal spines that are used in species identification.

SIZE

Brooks (1959) lists the size ranges of these small cladocerans as follows:

C. reticulata	♀ 0.6–1.4 mm	♂ 0.4–0.8 mm
C. lacustris	♀ 0.8–0.9 mm	♂ unknown
C. pulchella	♀ 0.4–0.7 mm	♂ to 0.5 mm
C. quadrangula	♀ to 1.0 mm	♂ to 0.6 mm
C. laticaudata	♀ to 1.0 mm	♂ to 0.7 mm

DISTRIBUTION AND ABUNDANCE

Most species of *Ceriodaphnia* are found throughout the United States and Canada (Brooks 1959). *C. quadrangula* is also found in Europe, Asia, and South America (Pennak 1978).

One or more of the *Ceriodaphnia* have been reported from each of the Great Lakes (Table 5). In our summer and fall samples the Duluth-Superior nearshore region of Lake Superior, we found *C. lacustris* present in low numbers. Lake Erie contains all five species. *C. lacustris* is the most limnetic of the five species, and peak abundances of 200–3500 m^{-3} have been reported from Lake Erie (Rolan et al. 1973; Watson and Carpenter 1974). This species is also

TABLE 5
Reports of *Ceriodaphnia* in the Great Lakes

	Sampling Date	Abd[a]	Reference		Sampling Date	Abd[a]	Reference
LAKE ERIE	1929	4F	Ewers 1933	**LAKE HURON**	1903, 1905, 1907	3, 5U	Bigelow 1922
	1938–1939	4U	Chandler 1940		1967–1968	5P	Patalas 1972
	unknown	2, 4U	Langlois 1954		1970	5R–P	Watson and Carpenter 1974
	1928–1929	1, 2, 3, 4F	Wilson 1960				
	1950–1951	4U	Davis 1954		1970–1971	5C, 3P	Carter 1972
	1966–1967	5U	Patalas 1972		1974	6P	Basch et al. 1980
	1967–1968	6U	Davis 1969		1974–1975	5P	McNaught et al. 1980
	1970	5P	Watson and Carpenter 1974				
	1971–1972	2, 4, 5U; 3R	Rolan et al. 1973	**LAKE ONTARIO**	1967	5P	Patalas 1969
	1973–1974	5P	Czaika 1978b		1967–1968	6P	Patalas 1972
					1970	5P	Watson and Carpenter 1974
LAKE MICHIGAN	1926	5U	Eddy 1927		1971–1972	5P	Wilson and Roff 1973
	1969–1970	5P	Gannon 1972a		1972	5P	McNaught and Buzzard 1973
	1969–1970	3, 5P	Gannon 1974				
	1971	5P	Howmiller and Beeton 1971		1972	5C	Czaika 1974a
					1972–1973	5U	Czaika 1978a
	1971–1972	5P	Beeton and Barker 1974				
	1972	3P	Stewart 1974	**LAKE SUPERIOR**	1967–1969	6U	Swain et al. 1970b
	1972–1977	3P	Evans et al. 1980		1970–1971	6U	Conway et al. 1973
	1973–1974	5R	Torke 1975		1979–1980	5U	This study

[a] Abundance Code
1 = *C. laticaudata*   2 = *C. pulchella*   3 = *C. quadrangula*   4 = *C. recticulata*   5 = *C. lacustris*   6 = *Ceriodaphnia* spp.
R = rare   U = uncommon   P = present   C = common   A = abundant   F = found in fish stomach contents

common in Lakes Ontario, Huron, and Michigan where abundance peaks of 80–10,000 organisms m^{-3} have been reported (Howmiller and Beeton 1971; Carter 1972; Watson and Carpenter 1974). *C. quadrangula* is found in low numbers in Lakes Ontario and Huron and in moderate numbers (up to 2000 m^{-3}) in Lake Michigan (Johnson 1972; Stewart 1974).

LIFE HISTORY IN THE GREAT LAKES

Life history information for this genus in the Great Lakes is very limited. *C. lacustris* is generally found only in the summer and fall, with abundance peaks occurring between August and October (Patalas 1969; Carter 1972; Gannon 1972a; Wilson and Roff 1973; Rolan et al. 1973; Czaika 1974a, 1978b; Watson and Carpenter 1974; Torke 1975). Dramatic fluctuations in abundance were observed in Lake Erie from July to October 1971 (Rolan et al. 1973) instead of the usual slow increase in abundance to a peak followed by a sharp decline in numbers.

Reproduction is parthenogenetic during the summer, with males and ephippial females appearing in September and October (Czaika 1974a). Like most cladocerans, this species overwinters as resting eggs.

ECOLOGY IN THE GREAT LAKES (AND OTHER AREAS)

Habitat. *Ceriodaphnia* have been found in both the littoral and limnetic zones of lakes (Brooks 1959). In the Great Lakes these organisms are usually found nearshore (Rolan et al. 1973) or in the warmer upper layers of the water column (Wilson and Roff 1973).

Diurnal Migrations. Wilson and Roff (1973) observed a limited vertical migration of *C. lacustris* in Lake Ontario. The mean depth of the population was closer to the surface after sunset than during the afternoon and evening. The mean population depth was not correlated with the intensity of the incident radiation at that depth. Juday (1904) and

McNaught and Haster (1966) were unable to define distinct migration patterns for the *Ceriodaphnia* populations they observed.

Feeding Ecology. *Ceriodaphnia*, like the other members of the family Daphnidae, are filter feeders (Brooks 1959). O'Brien and De Noyelles (1974) determined the filtering rate of *C. reticulata* in several ponds with different levels of phytoplankton. McNaught et al. (1980) found that Lake Huron *Ceriodaphnia* species filtered nannoplankton at a rate of 0.270 ml · animal^{-1} · hour^{-1} in Lake Huron.

As Food for Fish and Other Organisms. *Ceriodaphnia* are eaten by many species of fish including rock bass, largemouth bass, shiners, carp, mosquito fish, yellow perch, and crappie (Pearse 1921: Ewers 1933; Wilson 1960). Anderson (1970) also found that the copepods *Diacyclops thomasi* and *Acanthocyclops vernalis* consume *Ceriodaphnia quadrangula*.

Daphnia galeata Sars 1864 *mendotae* Birge 1918

TAXONOMIC HISTORY

This American *Daphnia* was first described by Smith (1874a). He identified it as *D. galeata* Sars 1864. S. A. Forbes (1882) decided that the American form more closely resembled *D. hyalina* Leydig 1860 and changed its name accordingly. This classification was used by most Great Lakes researchers until 1922, when Clemens and Bigelow reported *Daphnia longispina* Müller from Lake Huron. At that time *D. longispina* was a broadly defined taxon that included all *Daphnia* with uniformly small pectens on the postabdominal claw. During the next 30–40 years most *Daphnia* in Great Lakes collections were referred to as *D. longispina*. (The true identity of these organisms is unknown, but it is generally assumed that the majority were *D. galeata mendotae*.) Brooks (1957) concluded that the American species he examined were more closely related to *D. galeata* than to *D. hyalina*. To show the geographic substatus of these organisms, he renamed them with the trinomial *Daphnia galeata mendotae*. The name "*mendotae*" refers to a form of *D. longispina* var. *hyalina* Leydig 1860 figured in Birge (1918). All recent Great Lakes studies have used Brook's system of classification.

DESCRIPTION

The most detailed description of this species is presented by Brooks (1957). *Daphnia* can be distinguished from other cladocerans by their distinct rostrum and oval carapace that usually terminates in an elongated spine. *D. galeata mendotae* is distinguished from other *Daphnia* by its ocellus, broad pointed helmet (Plate 8), and the very fine, uniform pecten on the postabdominal claw (not visible at 50× magnification). Males are distinguished from females by their elongate first antennae (Plate 9).

Head shape is varible (Plate 10), but the peak of the helmet is generally near the midline of the body.

SIZE

Brooks (1959) reported adult females ranging from 1.3–3.0 mm long while males were much smaller, measuring only 1.0 mm. In Lake Superior we found females from 1.5–2.0 mm and males averaging 1.25 mm. The dry weight of animals from Lake Michigan varies from 2.5–8.9μg (Hawkins and Evans 1979).

DISTRIBUTION AND ABUNDANCE

D. galeata mendotae has been collected from lakes along the West Coast and in the formerly glaciated regions of the United States and Canada. It is also common in mountain lakes in Central and South America (Brooks 1957).

This cladoceran is found in all of the Great Lakes (Table 6). It is present in small numbers in lakes Ontario and Huron (Patalas 1969, 1972; Carter 1972; McNaught and Buzzard 1973), in moderate densities (100–6000 m^{-3}) in lakes Superior and Michigan (Wells 1970; Gannon 1974; Selgeby 1974; Stewart 1974; Watson and Wilson 1978), and is often quite abundant (average 900–5000 m^{-3}, peaks of 270000 m^{-3}) in Lake Erie (Bradshaw 1964; Davis 1968, 1969; Rolan et al. 1973).

LIFE HISTORY IN THE GREAT LAKES

Adults and juveniles are generally absent or present in very low numbers during the winter and spring months (Wells 1960; Gannon 1972a, 1975; Rolan et al. 1973; Stewart 1974; Beeton and Barker 1974; Selgeby 1975a; Heberger and Reynolds 1977). The animals hatch from ephippial eggs in early summer and reproduce rapidly, with the population size peaking in late summer or early fall (Davis 1961; Bradshaw 1964; Gannon 1972a, 1974, 1975; Rolan et al. 1973; Britt et al. 1973; Stewart 1974; Selgeby 1975a; Heberger and Reynolds 1977).

The life cycle of this species is best described by Selgeby (1975a), who observed three pulses each year in Lake Superior. The first generation hatches from ephippial eggs in late spring or early summer. The darkly pigmented, round-headed females reproduce parthenogenetically in

TABLE 6
Reports of *Daphnia galeata mendotae* in the Great Lakes

	Sampling Date	Abd[a]	Reference		Sampling Date	Abd[a]	Reference
LAKE ERIE	1919–1920	F	Clemens and Bigelow 1922	**LAKE HURON**	1903, 1905, 1907	A	Bigelow 1922
	1928–1929	CL	Wilson 1960		1907	U	Sars 1915
	1938–1939	AL	Chandler 1940		1919–1920	FL	Clemens and Bigelow 1922
	1949	A	Bradshaw 1964		1968	R	Patalas 1972
	1950–1951	CL	Davis 1954		1970–1971	P	Carter 1972
	1956–1957	CL	Davis 1962		1974	U	Basch et al. 1980
	1961	A	Britt et al. 1973		1974–1975	P	McNaught et al. 1980
	1962	RF	Wolfert 1965				
	1967	A	Davis 1962	**LAKE ONTARIO**	1919–1920	F	Clemens and Bigelow 1922
	1967–1968	A	Davis 1968		1927–1928	FL	Pritchard 1929
	1968	C	Patalas 1972		1967	U	Patalas 1969
	1968, 1970	C	Heberger and Reynolds 1977		1967	R	Patalas 1972
	1971–1972	C–A	Rolan et al. 1973		1969–1972	P	McNaught and Buzzard 1973
	1974–1975	C–A	Boucherle and Frederick 1976		1972	P	Czaika 1974a
					1972–1973	R	Czaika 1978a
LAKE MICHIGAN	1881	P	Forbes 1882				
	1893	P	Birge 1893	**LAKE SUPERIOR**	1871	—	Smith 1874b
	1887–1888	PL	Eddy 1927		1893	P	Birge 1893
	1954–1955, 1958	C	Wells 1960		1928	CL	Eddy 1934
	1954–1961	F	Wells and Beeton 1963		1933–1934	AL	Eddy 1943
	1966, 1968	U	Wells 1970		1960	PL	Putnam 1963
	1969–1970	P	Gannon 1975		1964	CL	Olson and Odlaug 1966
	1969–1970	C	Gannon 1974		1968	P	Patalas 1972
	1969–1970	P	Gannon 1972a		1971–1972	C	Selgeby 1975a
	1971	C	Howmiller and Beeton 1971		1971	C	Selgeby 1974
	1971–1972	U	Beeton and Barker 1974		1973	C	Upper Lakes Ref. Group 1977
	1973	C	Stewart 1974		1973	C	Watson and Wilson 1978
	1974	U	Evans and Stewart 1977		1974	P	Basch et al. 1980
	1972–1977	P	Evans et al. 1980		1979–1980	C–A	This study
	1975–1977	—	Hawkins and Evans 1979				

[a]Abundance Code
R = rare   U = uncommon   P = present   C = common   A = abundant   F = found in fish stomach contents
L = organisms classified as *D. longispina*   — = abundance ranking not appropriate

---

early July, producing clutches averaging 7 eggs that develop into low-helmeted females. This second generation also reproduces parthenogenetically between July and October. The clutches are somewhat smaller than those of the first generation, averaging 5.7 eggs. Strongly helmeted females and numerous males hatch from these eggs. This third generation reproduces sexually in the fall and produces ephippial eggs that overwinter.

Males have been observed in October and November in Lakes Ontario and Michigan (Czaika 1974a; Stewart 1974) and may compose 30% of the population late in the year.

ECOLOGY IN THE GREAT LAKES

Habitat. *D. galeata mendotae* is found in many lakes but is most common in large, deep, transparent waters (Patalas 1971). It prefers the upper water strata (Brooks 1959) several meters below the surface during the day (Wells 1960; Heberger and Reynolds 1977). In Lake Michigan *D. galeata mendotae* is often distributed higher in the water column than *D. retrocurva* (Stewart 1974).

D. galeata mendotae has also been found near the bottom in Lake Michigan (Wells 1960). Greater concentrations occasionally occur in bottom cores from shallow water than in the overlying water column (Evans and Stewart 1977). Heberger and Reynolds (1977) found this species well adapted to a benthic life, being able to withstand very low levels of dissolved oxygen that often occur in the hypolimnion.

This species is often found in large swarms (up to 2×10^7 m^{-3}) due to the action of the wind and waves (Boucherle and Frederick 1976). These swarms tend to disperse rapidly.

Diurnal Migrations. The maximum concentration is often found at a depth of 10 m during the day (Wells 1960). Juveniles are usually located slightly above the adults in the water column and arrive at the surface first when the upward migration begins 1.5 hours after sunset. Many of the animals move down slightly after midnight but return to the surface before settling to their daytime depths an hour before sunrise.

Feeding Ecology. *Daphnia* are filter feeders that consume small algae (0.8–40.0μm) and prefer chlorophytes over other algal types (Burns 1969; Porter 1977). McNaught et al. (1980) studied the filtering rate, ingestion rate, and assimilation rate of *D. galeata mendotae* feeding on both nannoplankton and net plankton in southern Lake Huron.

As Food for Fish. *D. galeata mendotae* is occasionally consumed by bloater and cisco (Clemens and Bigelow 1922; Wells and Beeton 1963). In Lake Superior it is often the only food item found in the stomachs of juvenile smelt in late summer (Selgeby 1974).

LIFE HISTORY AND ECOLOGY IN OTHER LAKES

Hall (1964) conducted laboratory studies on the reproductive cycle of *D. galeata mendotae* from Base Line Lake where it is present year-round and males appear twice each year. He found that each female produces 2 to 4 eggs per clutch. Three developmental stages are recognizable in the embryos before they are released from the brood chamber. Egg development time is temperature dependent, varying from 2 days at 25°C to 20 days at 4°C. Adult life span is also related to temperature, lasting 150 days at 11°C and only 30 days at 25°C. The growth rate of this species also depends on the available food supply.

Other studies (Burns 1969) have shown that *Daphnia*'s filtering rate increases 250% as water temperature is raised from 15 to 20°C. A slight additional increase in filtering rate is observed at 25°C.

In other populations of *D. galeata mendotae*, Haney and Hall (1975) observed vertical migration patterns similar to those in the Great Lakes. In addition, they observed higher filtering rates at the surface at night than at the noon depths, suggesting that vertical migration may be advantageous to feeding.

Hall (1964) reprots that this species of *Daphnia* is very tolerant to variations in light and alkalinity. In later studies Hensey and Porter (1977) found that, although *D. galeata mendotae* prefers high dissolved oxygen levels, it has a broad tolerance for this factor and occasionally occurs in oxygen-poor regions near the thermocline.

Daphnia retrocurva Forbes 1882

TAXONOMIC HISTORY

The taxonomy of this species has undergone several changes since it was first described by S. A. Forbes in 1882. Herrick (1884) and Birge (1893) recognized the similarity of *Daphnia retrocurva* to a previously described species and therefore synonymized *D. retrocurva* with *D. kahlbergensis*. Differences were noted between the Great Lakes specimens and the description of the type specimen, leading to the report of several variations including *D. kahlbergensis* var. *intexta* Birge 1893, *D. kahlbergensis* var. *retrocurva* Birge 1893 and *D. retrocurva* var. *intexta* Forbes 1891.

Later work by Richard (1896) and Sars (1915) placed *retrocurva* as a distinct species in the new genus *Hyalodaphnia*, which included all *Daphnia* lacking ocelli. It was at that time that *Hyalodaphnia retrocurva* var. *intexta* was reported from Lake Huron (Sars 1915). The genus *Hyalodaphnia* was abandoned when it was found to contain several heterogeneous species, and the name *Daphnia retrocurva* came back into use (Birge 1918).

Later, Woltereck (1932) suggested that the term "retrocurva" was misleading because many individuals of this species lacked the recurved helmet. He renamed this cladoceran *D. pulex* var. *parapulex*. Many other researchers did not agree with him, and Brooks (1946) reestablished *D. retrocurva* as the proper name for this species.

DESCRIPTION

D. retrocurva is distinguished from the other Great Lakes *Daphnia* by its lack of an ocellus and its recurved helmet with its peak located dorsal to the midline of the body (Plates 11, 12). With higher magnification the enlarged middle pecten of the postabdominal claw can be observed. *D. retrocurva* undergoes cyclomorphosis and varies a great deal in its helmet shape (Plate 13). Stewart (1974) observed the greatest helmet development in populations from Lake

Erie, but all Great Lakes specimens have less recurved helmets than those from smaller lakes. A more detailed description of this species is given by Brooks (1957, 1959).

SIZE

Brooks (1959) reported the length of this species (including helmet) as follows: ♀ 1.3–1.8 mm, ♂ 1.0 mm. In Lake Superior we found animals with a similar size range. The dry weight of Lake Michigan specimens varies seasonally from 1.2–6.5 μg, with the heaviest animals occurring in the fall (Hawkins and Evans 1979).

DISTRIBUTION AND ABUNDANCE

D. retrocurva is generally found east of the Rocky Mountains in the formerly glaciated areas of the northern United States and southern Canada. It has also been collected from lakes in the Puget Sound region.

This species is one of the most abundant cladocerans in the Great Lakes and has been reported from all five lakes since the early 1900s (Table 7). Numbers are highest in the summer with maximum densities of 2000–24000 organisms m^{-3} occurring in Lake Michigan (Wells 1970; Howmiller and Beeton 1971; Gannon 1974; and Stewart 1974);

TABLE 7
Reports of *Daphnia retrocurva* in the Great Lakes

	Sampling Date	Abd[a]	Reference		Sampling Date	Abd[a]	Reference
LAKE ERIE	1919–1920	F	Wickliff 1920	**LAKE MICHIGAN (continued)**	1973 1974	C	Torke 1975
	1919–1920	F	Clemens and Bigelow 1922		1974	P	Evans and Stewart 1977
	1928–1930	C	Wright 1955		1972–1977	P	Evans et al. 1980
	1929	F	Ewers 1933		1975–1977	P	Hawkins and Evans 1979
	1929	F	Kinney 1950				
	1938–1939	P	Chandler 1940	**LAKE HURON**	1903, 1905, 1907	A	Bigelow 1922
	1948–1949, 1959	A	Bradshaw 1964		1907	C	Sars 1915
	1950–1951	A	Davis 1954		1968	P	Patalas 1972
	1956–1957	A	Davis 1962		1970–1971	C	Carter 1972
	1961	C	Britt et al. 1973		1974	C	Basch et al. 1980
	1962	F	Wolfert 1965		1974–1975	C	McNaught et al. 1980
	1967	P	Davis 1968				
	1967–1968	C	Davis 1969	**LAKE ONTARIO**	1919–1920	F	Clemens and Bigelow 1922
	1968	C	Patalas 1972		1967–1968	C	Patalas 1969
	1968, 1970	C	Heberger and Reynolds 1977		1967	C	Patalas 1972
	1971–1973	A	Rolan et al. 1973		1969, 1972	P	McNaught and Buzzard 1973
	1974–1975	C–A	Boucherle and Frederick 1976		1971–1972	C	Wilson and Roff 1973
					1972	C	Czaika 1974a
LAKE MICHIGAN	1887–1888	C	Eddy 1927		1972, 1973	R–P	Czaika 1978a
	1926–1927	P	Eddy 1934				
	1954–1955, 1958	C	Wells 1960	**LAKE SUPERIOR**	1889	A	Forbes 1891
	1954–1961	F	Wells and Beeton 1963		1893	P	Birge 1893
	1964–1965	P	McNaught and Hasler 1966		1928	P	Eddy 1934
	1966, 1968	U–C	Wells 1970		1968	P	Patalas 1972
	1969–1970	A	Gannon 1974		1971	A	Selgeby 1974
	1969–1970	C	Gannon 1975		1971–1972	C	Selgeby 1975
	1969–1970	P	Gannon 1972a		1973	C	Upper Lakes Ref. Group 1977
	1971	C	Howmiller and Beeton 1971		1973	P	Watson and Wilson 1979
	1973	C	Stewart 1974		1974	C	Basch et al. 1980
					1979–1980	C	This study

[a] Abundance Code
R = rare   U = uncommon   P = present   C = common   A = abundant   F = found in fish stomach contents

4000–10000 m⁻³ in Lake Erie (Davis 1954; Bradshaw 1964); 1300–2500 m⁻³ in Lake Superior (Selgeby 1974, 1975a; this study); and 1200 m⁻³ in Lake Ontario (McNaught and Buzzard 1973). Average yearly densities ranging from 35–1200 organisms m⁻³ have been reported from lakes Superior (Selgeby 1974) and Erie (Davis 1954), respectively. Because of its great abundance and relatively large size, *D. retrocurva* comprises a considerable proportion of the total summer biomass of crustacean zooplankton (Davis 1962).

LIFE HISTORY IN THE GREAT LAKES

The hatching of ephippial eggs of *D. retrocurva* may be temperature dependent. Immature animals have first been collected in April (Wright 1955) and May (Rolan et al. 1973) in Lake Erie and in June (Wells 1960; Patalas 1969; Wilson and Roff 1973; Selgeby 1974, 1975a) and July (Gannon 1972a; Czaika 1974a; Stewart 1974; Torke 1975) in the other lakes. This wide range of hatching times may be due to the great variation in spring water temperatures in the Great Lakes, with the shallowest areas (Erie) warming first. Numbers increase rapidly after hatching with abundance peaks occurring as early as June in Lake Erie (Davis 1954; Bradshaw 1964; Heberger and Reynolds 1977) and from July to September in the other lakes (Wells 1960; Patalas 1969; Gannon 1972a, 1975; Wilson and Roff 1973; Czaika 1974a; Stewart 1974; Torke 1975; Selgeby 1975a). Abundance declines rapidly in October and November, with very few adults or immatures reported from December zooplankton collections.

Adult females appear between June and July and reproduce parthenogenetically throughout the summer. The first few generations have low helmets (Stewart 1974; Torke 1975; Selgeby 1975a). Helmet size increases in the summer and fall generations. Sexual animals are produced in late fall and exhibit highly developed recurved helmets. These males and females mate between late August and November. The females then produce the ephippial eggs that overwinter (Davis 1962; Gannon 1972a, 1974; Stewart 1974; Czaika 1974a; Selgeby 1974, 1975a).

ECOLOGY IN THE GREAT LAKES

Habitat. *D. retrocurva* is most common in the nearshore regions of the Great Lakes (Carter 1972; Heberger and Reynolds 1977; Watson and Wilson 1978), appearing in the open water zone only during peak abundance. Huge swarms have been observed in Lake Erie (Boucherle and Frederick 1976) due to the action of wind and waves and changes in lake level. These swarms disperse quite rapidly.

Diurnal Migration. *D. retrocurva* has a well-defined migration pattern similar to that of *D. galeata mendotae* (Wells 1960; McNaught and Hasler 1966). During the day the animals are located below the thermocline but by late afternoon begin to migrate to the surface at rates of up to 10.6 m/hr. The young reach the surface first with the adults joining them soon after sunset. They remain near the surface during the night and at sunrise begin their descent at rate of 5 m/hr. Movement is positively correlated with the times of most rapid changes in light intensity in the water column.

Feeding Ecology. Birge (1897) studied the food habits of this filter feeder in Lake Mendota. He found that *D. retrocurva* consumed mainly diatoms and *Anabaena*. In Lake Huron the filtering, ingestin, and assimilation rates have been determined for *Daphnia* feeding on nannoplankton and net plankton (McNaught et al. 1980).

As Food for Fish and Other Organisms. Small bass, bloaters, cisco, crappie, logperch, sauger, freshwater drum, shiners, troutperch, and yellow perch consume *D. retrocurva* in the Great Lakes (Wickliff 1920; Clemens and Bigelow 1922; Ewers 1933; Kinney 1950; Wells and Beeton 1963). Predation experiments (Lane 1978) showed that *D. retrocurva* is also eaten by other zooplankton, including *Diacyclops thomasi*, *Leptodora kindti*, *Tropocyclops prasinus mexicanus*, *Chaoborus*, *Mesocyclops edax*, and *Skistodiaptomus oregonensis*.

LIFE HISTORY AND ECOLOGY IN OTHER LAKES

Patalas (1971) found that *D. retrocurva* was most common in lakes of intermediate trophic status. Its life cycle in most lakes is similar to that in the Great Lakes: immature forms appear in the spring, numbers increase rapidly during the summer, then decline in late fall after the appearance of males and ephippial females. The floating ephippia are often blown inshore, and the eggs hatch in the littoral zone during the next spring (Forbes 1891; Birge 1897; Hall 1964; Hutchinson 1967).

Daphnia longiremis (Sars 1861)

TAXONOMIC HISTORY

Although Sars (1862) described *Daphnia longiremis* as a distinct species, other researchers considered it a subspecies of the previously described *D. cristata*. Friedenfelt (1913) examined the *Daphnia* from lakes that contained

TABLE 8
Reports of *Daphnia longiremis* in the Great Lakes

	Sampling Date	Abd[a]	Reference		Sampling Date	Abd[a]	Reference
LAKE ERIE	1956–1957	P	Davis 1962	**LAKE HURON**	1968	P	Patalas 1972
	1968	P	Patalas 1972		1970	C	Carter 1972
	1971–1973	P	Rolan et al. 1973		1974	P	Basch et al. 1980
	1968, 1970	U	Heberger and Reynolds 1977		1974–1975	U	McNaught et al. 1980
				LAKE ONTARIO	1967–1968	U	Patalas 1969
LAKE MICHIGAN	1966, 1968	U	Wells 1970		1967	P	Patalas 1972
	1969–1970	P	Gannon 1972a		1972	R	Czaika 1974a
	1969–1970	R	Gannon 1975		1972–1973	R	Czaika 1978a
	1969–1970	P	Gannon 1974				
	1971	U–P	Howmiller and Beeton 1971	**LAKE SUPERIOR**	1973	P	Upper Lakes Ref. Group 1977
	1971–1972	U	Beeton and Barker 1974		1974	P	Basch et al. 1980
	1972–1977	P	Evans et al. 1980				
	1973	U	Stewart 1974				

[a]Abundance Code

R = rare   U = uncommon   P = present   C = common   A = abundant   F = found in fish stomach contents

both "species." He found no evidence of transitional morphological forms and concluded that *Daphnia longiremis* was indeed a separate species.

During the mid-1900s most *Daphnia* possessing uniformly small pecten on the postabdominal claw were classified as *D. longispina*. *Daphnia longiremis* was often placed in this group as a subspecies (Kiser 1950). Later taxonomists discarded the heterogeneous *longispina* grouping and *Daphnia longiremis* reappeared in the literature as a distinct species (Brooks 1957).

DESCRIPTION

D. longiremis is distinguished from other *Daphnia* species by the long "hairs" on the second antennae that reach to the posterior margin of the carapace when straightened (Plate 14). The setae located at the base of the second segment of the three-segmented branch of the second antenna does not reach to the end of that branch. There is a prominent space between the rostrum and the body of the animal. An ocellus is absent, and the uniformly fine pecten of the postabdominal claw are not readily visible under low magnification. The shape of the helmet and body form are quite variable (Plate 15).

As in other species of *Daphnia*, males are recognized by their elongate first antennae. Male *D. longiremis* are extremely rare.

Brooks (1957, 1959) gives a more detailed description of this cladoceran.

SIZE

Females ranging from 0.6–1.2 mm long have been reported (Brooks 1959). Females from Lake Erie (present study) averaged 1.07 mm and ranged from 0.2–1.2 mm.

DISTRIBUTION AND ABUNDANCE

D. longiremis is a holarctic, cold water, stenothermic species found in the northern United States, Canada, northern Europe, and Asia (Brooks 1957). It has only recently been reported from the Great Lakes (Table 8). Davis (1962) suspects that this cladoceran was present in earlier zooplankton collections but was classified as *Daphnia longispina*.

This species is most abundant in inshore areas of Lake Michigan's Green Bay (Howmiller and Beeton 1971; Gannon 1974) and Lake Erie's Cleveland Harbor area (Rolan et al. 1973). Maximum desities of 200–2700 and 1700 organisms m[-3] have been reported from these areas, respectively. In the offshore regions of Lake Michigan, Wells (1970) found much lower densities of *D. longiremis* (< 6 m[-3]). This species is also found in low numbers in Lakes Huron, Ontario, and Superior.

LIFE HISTORY IN THE GREAT LAKES

In the Great Lakes *D. longiremis* is present in low numbers year-round (Rolan et al. 1973; Gannon 1974), with maximum densities occurring during the summer (Davis 1962;

Wells 1970; Carter 1972; Gannon 1974; Heberger and Reynolds 1977). In Cleveland Harbor, abundance was highest in November (Rolan et al. 1973).

Reproduction is believed to be exclusively parthenogenetic, as very few males or ephippial females have been reported (Brooks 1957).

ECOLOGY IN THE GREAT LAKES

Habitat. The Great Lakes region is at the southern edge of the distribution range of this lake-dwelling species (Brooks 1957). During stratified conditions it is usually found in the cool waters below the thermocline but above the poorly oxygenated regions of the hypolimnion (Howmiller and Beeton 1971). Heberger and Reynolds (1977) observed that *D. longiremis* disappeared from Lake Erie at the onset of oxygen depletion in this lake.

Diurnal Migration. The vertical migration pattern of this species is not well defined. The only time that Wells (1970) observed *D. longiremis* moving towards the surface of Lake Michigan at night was in June when the surface water was less than 11°C. Only a small proportion of the population was involved in this migration.

Feeding Ecology and As Food For Fish. This species filter feeds on both nannoplankton and net plankton in Lake Huron. Filtering rates of 0.217 ml $\cdot$ animal^{-1} $\cdot$ hour^{-1} have been measured (McNaught et al. 1980).

D. longiremis has not yet been reported as a prey item of any species of fish in the Great Lakes.

Family Bosminidae

TAXONOMIC HISTORY

The taxonomy of this family has undergone several revisions since the first bosminid, *Lynceus longirostris*, was described in 1785 by O. F. Müller. Shortly after this report Jurine (1820) described another bosminid, giving it the name *Monoculus cornutus*. Baird (1845) considered these animals with "pendulate antennules" dinstinct from the family Daphnidae and established the family Bosminidae with the genue *Lynceus*. The genus name was later changed to *Bosmina*, and another species, *B. coregoni*, was added (Baird 1857).

The number of described species in this family grew rapidly until 1900. Species were distinguished by antennal length and number of segments, body shape and size, and the length of the mucro (a projection from the posterior edge of the carapace). Since these characters show regional variation and undergo cyclomorphosis, many of the described species were synonymous. Burkhardt (1900) discovered that the morphology of the postabdominal claw was not variable and used this characteristic to reclassify the 56 reported species into two species: *Bosmina longirostris* and *B. coregoni*.

Frey (1962) and Goulden and Frey (1963) reported that species of *Bosmina* could also be distinguished by pores in their head shields. They reported seven species of *Bosmina* with two, *B. longirostris* and *B. coregoni*, occurring in North America.

Further work by Deevey and Deevey (1971) used the morphology of the postabdominal claw, head pores, antennules, mucro, shell, and clasper of the male's first leg to identify five species of bosminids in North America. The former members of the "coregoni" group were placed in the genus *Eubosmina* and split into four species: *E. coregoni*, *E. longispina*, *E. hagmanni*, and *E. tubicen*. *Bosmina longirostris* remained the sole American species of the genus *Bosmina*.

Most of the characteristics used by Deevey and Deevey to distinguish species are not readily observed with a dissecting microscope. Only one species, *E. coregoni*, is easily recognized due to its lack of a mucro. (The other four species possess mucrones).

In our samples from Lake Superior we only found two species of bosminids, *E. coregoni* and *B. longirostris*. (Identification was aided by examination of the postabdomen and first leg of males collected in the fall of 1979). However, there may be other species of the mucronate bosminids in the littoral regions of the Great Lakes.

Bosmina longirostris (O. F. Müller) 1785

TAXONOMIC HISTORY

The taxonomy of this species is discussed under the family Bosminidae. In this section, we include information from all reports of *B. longirostris*, even though some may have been referring to other species of mucronate bosminids that had not yet been described. Early reports of *Eubosmina longispina* may refer to a long-featured morph of *B. longirostris* (Torke 1975).

DESCRIPTION

The antennules of the female are fairly large and fixed to the head, curving backward parallel to each other (Plate 16). The posterior margin of the carapace extends into a pointed mucro. Under high magnification the proximal and distal pecten on the postabdominal claw and the sensory bristle (located midway between the eye and the base of the antennules) can be observed. Males are distinguished from females by their smaller size, notched antennules, and modified first legs (Plate 17). Cyclomorphosis is not common.

More detailed descriptions are available in Deevey and Deevey (1971).

SIZE

In Lake Superior we found mature females from 0.4–0.6 mm long with an average size of 0.46 mm. Males were slightly smaller, averaging 0.43 mm and ranging from 0.4–0.5 mm. In Lake Michigan (Stewart 1974) females vary in length from 0.2–0.6 mm and have a dry weight of 0.6–1.8μg (Hawkins and Evans 1979). The summer animals are the smallest.

DISTRIBUTION AND ABUNDANCE

B. longirostris has a worldwide distribution and has been found in both lakes and ponds. It has been reported from Lakes Erie and Superior for over a hundred years (Table 9) and is one of the most abundant crustaceans during the summer. Abundance peaks of 2000–142000 organisms m^{-3} have

TABLE 9
Reports of *Bosmina longirostris* in the Great Lakes

	Sampling Date	Abd[a]	Reference		Sampling Date	Abd[a]	Reference
LAKE ERIE	1880, 1882	U	Vorce 1881	**LAKE HURON**	1903, 1905, 1907	C	Bigelow 1922
	1919–1920	U	Clemens and Bigelow 1922		1907	P	Sars 1915
	1928–1930	U	Wright 1955		1970	C	Watson and Carpenter 1974
	1928	F	Sibley 1929				
	1928–1929	P	Wilson 1960		1974	C	Basch et al. 1980
	1929	F	Ewers 1933		1974–1975	C–A	McNaught et al. 1980
	1938–1939	P	Chandler 1940				
	1948–1949, 1959	P	Bradshaw 1964	**LAKE ONTARIO**	1919–1920	U	Clemens and Bigelow 1922
	1950–1951	C	Davis 1954		1927–1928	F	Pritchard 1929
	1956–1957	C	Davis 1962		1939–1940	P	Tressler et al. 1953
	1961	C	Britt et al. 1973		1967	A	Patalas 1969
	1961–1962	PF	Hohn 1966		1970–1971	A	Carter 1972
	1967–1968	C	Davis 1962		1970	C–A	Watson and Carpenter 1974
	1968–1969	P	Heberger and Reynolds 1977		1972	A	McNaught and Buzzard 1973
	1970	P	Watson and Carpenter 1974		1972	C–A	Czaika 1974a
	1971–1972	A	Rolan et al. 1973		1973	P	Czaika 1978a
LAKE MICHIGAN	1954–1955, 1958	P	Wells 1960	**LAKE SUPERIOR**	1889	P	Forbes 1891
	1954–1961	PF	Wells and Beeton 1963		1893	P	Birge 1893
	1966, 1968	P	Wells 1970		1913	F	Hankinson 1914
	1969–1970	C	Gannon 1975		1928	P	Eddy 1934
	1969–1970	A	Gannon 1974		1967–1969	U–P	Swain et al. 1970b
	1969–1970	C–A	Gannon 1972a		1971–1972	A	Selgeby 1975a
	1971	C	Howmiller and Beeton 1971		1971	C	Selgeby 1974
	1972–1977	P	Evans et al. 1980		1973	C	Upper Lakes Ref. Group 1977
	1973	A	Stewart 1974		1973	C	Watson and Wilson 1978
	1973–1974	C–A	Torke 1975		1974	C	Basch et al. 1980
	1974	P	Evans and Stewart 1977		1979–1980	C–A	This study
	1975–1977	—	Hawkins and Evans 1979				

[a]Abundance Code
R = rare   U = uncommon   P = present   C = common   A = abundant   F = found in fish stomach contents
— = abundance ranking not appropriate

been found in Lake Huron (Watson and Carpenter 1974; McNaught et al. 1980), 4700 in Lake Erie (Bradshaw 1964), 18000 in Lake Ontario (Watson and Carpenter 1974), and from 29000–230000 in Lake Michigan (Gannon 1974; Stewart 1974).

In Lake Superior Selgeby (1974, 1975a) reported yearly averages of 308 and 918 m^{-3} from sites in the Apostle Islands and 187 m^{-3} at Sault Ste. Marie. We found the highest numbers at littoral stations at the western tip of the lake. Peaks of less than 1000 m^{-3} were observed offshore, 13000 m^{-3} in shallow water and 42000 m^{-3} in the Duluth-Superior Harbor.

LIFE HISTORY IN THE GREAT LAKES

The bosminids are opportunistic species like the daphnids. They overwinter mainly as resting eggs, and when conditions become favorable in the spring they reproduce rapidly by parthenogenesis. Abundance generally peaks in late summer or early fall (Wilson 1960; Watson and Wilson 1978). Numbers decrease rapidly as the food supply decreases in the fall but may remain high until December in nutrient-rich areas such as the Milwaukee Harbor (Gannon 1972a).

Selgeby (1974, 1975a) followed the reproductive cycle of *B. longirostris* in Lake Superior. It was one of the earliest reproducing cladocerans, with females carrying 9–12 eggs in early June. By the end of the month the average clutch size declined to 4.5 eggs per female. The population increased seven-fold by mid-July, and, although reproduction continued, mean clutch size was only 3.5. During August and October the population and clutch size both declined. Males and ephippial females appeared in October and produced wintering eggs. Similar reproductive patterns were observed in this study (1979–1980), by Stewart (1974), Torke (1975), and Gannon (1972a), although some seasonal variations were noted.

ECOLOGY IN THE GREAT LAKES

Habitat. *B. longirostris* prefers cool, well-oxygenated waters (Bhajan and Hynes 1972) and littoral areas (Wilson 1960; Czaika 1974; Evans and Stewart 1977).

In 1923 Gurney found that this species was more common in "silty" or eutrophic lakes. Beeton (1965) also observed that *B. longirostris* tends to replace *E. coregoni* as bodies of water become more eutrophic. However, due to taxonomic problems in the genus *Bosmina* at that time, this "species" shift is doubtful and bosminids are not considered good trophic indicators at the current time (Gannon and Stemberger 1978).

Diurnal Migration. This species is often most numerous in the upper water layers (Wells 1960; Wilson and Roff 1973). In Lake Michigan, Wells found that a few bosminids moved toward the surface late in the day. A slight sinking was noted during the middle of the night, and a brief upward migration occurred near dawn.

Feeding Ecology. *B. longirostris* is a filter feeder consuming algae and protozoans in the size range of 1–3μm (Markarewicz and Likens 1975). McNaught et al. (1980) determined the filtering, ingestion, and assimilation rates for this species in southern Lake Huron. It appears to be a less effective competitor than *Daphnia*. In Lake Michigan *Bosmina* declined in abundance in August when the *Daphnia* population was peaking and food was becoming limiting (Stewart 1974). DeCosta and Janicki (1978) observed that *B. longirostris* may have a competitive advantage over other cladocerans in acidic waters (pH equals 4.6).

As Food for Fish and Other Organisms. This species has been found in small numbers in the diets of mosquitofish, bass, perch, crappie, walleye fry, and small bloater (Pearse 1921; Sibley 1929; Ewers 1933; Wells and Beeton 1963; Hohn 1966). It is a more important food for young whitefish in Lake Superior (Hankinson 1914) and for alewife in Lake Ontario (Pritchard 1929). *Mysis relicta*, the opossum shrimp, also consumes bosminids (Forbes 1882).

LIFE HISTORY AND ECOLOGY IN OTHER LAKES

B. longirostris is most abundant in the epilimnion and littoral areas of lakes where other species of *Bosmina* also occur (Holway and Cocking 1953; Hutchinson 1967).

B. longirostris responds to local conditions of temperature and food abundance that determine its monocyclic or dicyclic reproductive pattern (Zhdanova 1969; Schindler and Novan 1971). Dicyclic populations have spring and fall population peaks with the adults overwintering (Janicki and DeCosta 1977).

Water temperature, nutrition, and genetic constitution have been shown to control this species growth rate and fecundity (Bhajan and Hynes 1972). Zhdanova (1969) found best growth from 11 to 19°C in labortory cultures. Each female produced 5–8 clutches of 2–4 eggs each. Embryonic development took 2–3 days and sexual maturity was reached 3–4 days later with individuals living 20–25 days. Bhajan and Hynes (1972) observed better reproduction when cultures were maintained at temperatures less than 11°C.

Population size may be affected by oxygen depletion, especially when the upper water layers are affected (Zhdanova 1969).

Kerfoot (1974, 1977, 1978) studied *B. longirostris* populations in Lake Washington. He observed changes in egg size, body size, and body shape that may be related to predation by copepods and fish. He proposes that the survival of this species depends on its evolution of behavioral responses and morphological defenses that frustrate predators after the initial contact.

McNaught et al. (1980) felt that *B. longirostris* in Lake Huron was less susceptible to fish predation than *E. coregoni*. Stenson (1976) observed that predation played a role in relative abundances of both species in his study areas.

Eubosmina coregoni (Baird) 1857

TAXONOMIC HISTORY

The taxonomic history of this species is covered under the family Bosminidae. Although several reports of *Bosmina coregoni* and *Bosmina coregoni coregoni* appear in the Great Lakes zooplankton literature, we only include those that can be validated as *Eubosmina coregoni* by lacking a mucro.

DESCRIPTION

In both sexes the carapace of *E. coregoni* does not extend into a mucro although a slight point may be noted at the posterior margin. High power magnification is needed to observe the presence of only a proximal pecten on the postabdominal claw and the sensory bristle, which is located near the tip of the rostrum where the antennule attaches. The pores of the head shield are conspicuous and are located on the side of the head above the mandibles. The antennules of female *E. coregoni* are fixed to the head and are generally longer than those of *Bosmina longirostris* (Plate 18). Males are recognized by their notched antennules and modified first legs (Plate 19).

Deevey and Deevey (1971) give a more detailed description of this species.

SIZE

Deevey and Deevey (1971) report the average length of females as 0.60 mm and of males as 0.45 mm. In Lake Superior we found females ranging from 0.5–0.6 mm and males from 0.5–0.6 mm. In Lake Michigan *E. coregoni* has a dry weight of 1.2–2.5 µg (Hawkins and Evans 1979).

DISTRIBUTION AND ABUNDANCE

This is a common European species that has only recently been observed in the Great Lakes and smaller waters in Pennsylvania and Ontario, Canada.

In the Great Lakes *E. coregoni* is often one of the most abundant summertime species (Table 10). Abundance peaks reach 47000 and 15000 adults m^{-3} in lakes Erie and Michigan, respectively (Davis 1968; Howmiller and Beeton 1971; Rolan et al. 1973; Gannon 1974). In lakes Ontario and Huron numbers are generally less than 2000 m^{-3} (McNaught and Buzzard 1973; Watson and Carpenter 1974) with occasional peaks of 200000 m^{-3} occurring in inshore regions (McNaught et al. 1980). Although Putnam (1963), Olson and Odlaug (1966), and Swain et al. (1970b) reported this species from Lake Superior, the identity of their specimens has not been confirmed by the key subsequently published by Deevey and Deevey. *E. coregoni* occurred at low densities in western Lake Superior in 1971 (Selgeby 1975b) and at 2 of the 80 stations sampled during the lakewide survey of 1973 (Watson and Wilson 1978). It now seems to be common (100–2000 m^{-3}) during the spring and fall at the extreme western end of the lake and in the Duluth-Superior harbor (this study 1979–1980). In general, *E. coregoni* is less abundant than *B. longirostris*, although it may be the more common bosminid in fall samples (Stewart 1974; Torke 1975).

LIFE HISTORY IN THE GREAT LAKES

The reproductive pattern of this species is basically the same as that of *B. longirostris*. *E. coregoni* is absent or present in very low numbers during the winter and spring (Davis 1969; Wells 1970; Beeton and Barker 1974). Females hatch from overwintering resting eggs in the spring and begin reproducing parthenogenetically. The population increases rapidly, with abundance peaks generally occurring between September and November. Population peaks are often larger and occur earlier in areas that warm faster and stay warm later in the year (Gannon 1972a; Watson and Carpenter 1974; Czaika 1978a). Males appear in October and November and mate with females that will produce ephippial resting eggs to carry the species through the winter (Carter 1969; Gannon 1972a; Stewart 1974).

ECOLOGY IN THE GREAT LAKES

Habitat. *E. coregoni* is found at all depths, but it appears to prefer warm epilimnetic waters and occasionally concentrates at a depth of 10 m (Wells 1970). This species may be

TABLE 10
Reports of *Eubosmina coregoni* in the Great Lakes

	Sampling Date	Abd[a]	Reference		Sampling Date	Abd[a]	Reference
LAKE ERIE	1967	A	Davis 1968	**LAKE MICHIGAN (continued)**	1973–1974	P	Torke 1975
	1967–1968	C	Davis 1969		1974	P	Evans and Stewart 1977
	1968, 1970	A	Heberger and Reynolds 1977		1975–1977	—	Hawkins and Evans 1979
	1970	C	Watson and Carpenter 1974	**LAKE HURON**	1970	P	Watson and Carpenter 1974
	1971–1972	C–A	Rolan et al. 1973		1971–1972	U	Carter 1972
LAKE MICHIGAN	1966, 1968	P	Wells 1970		1974	U	Basch et al. 1980
	1969–1970	P–C	Gannon 1972a		1974–1975	C–A	McNaught et al. 1980
	1969–1970	A	Gannon 1974	**LAKE ONTARIO**	1970	C	Watson and Carpenter 1974
	1969–1970	P	Gannon 1975		1972	P–C	Czaika 1974a
	1971	C	Howmiller and Beeton 1971		1972–1973	C	Czaika 1978a
	1971–1972	P	Beeton and Barker 1974	**LAKE SUPERIOR**	1971	U	Selgeby 1975a
	1972–1977	P	Evans et al. 1980		1973	U	Upper Lakes Ref. Group 1977
	1973	P	Stewart 1974		1974	P	Basch et al. 1980
					1979–1980	P–C	This study

[a] Abundance Code
R = rare   U = uncommon   P = present   C = common   A = abundant   F = found in fish stomach contents
— = abundance ranking not appropriate

less tolerant to eutrophication than *B. longirostris*, as it is mainly found in larger, less productive lakes (Gurney 1923).

Diurnal migration.   No studies of the migration patterns of this species have been reported from the Great Lakes. Hutchinson (1967) observed that in other lakes immature forms migrated while the adults did not. The animals came to the surface before dark at rates of 3.8–13.2 cm/hr and began their descent an hour before dawn at a rate of 22.2 cm/hr.

Feeding Ecology and As Food for Fish.   *E. coregoni* has only been reported from the Great Lakes since 1966 (Wells 1970), and little is known about its role in the food chain. Like the other bosminids, it is a filter feeder. The filtering, ingestion, and assimilation rates of this species have been determined in southern Lake Huron by McNaught et al. (1980). These authors suggest that *E. coregoni* may be more susceptible to fish predation in Lake Huron than the slightly smaller *Bosmina longirostris*.

LIFE HISTORY AND ECOLOGY IN OTHER LAKES

Zhdanova (1969) studied *E. coregoni* in fresh-water lakes and ponds in Kiev, Russia. He observed parthenogenetic reproduction from June through August when the water temperature ranged from 18 to 23°C. Maximum abundances of 223000–292000 organisms m^{-3} were noted. Males and ephippial females appeared in late September when water temperature declined to 15°C. By November, when the water had cooled to 4°C, this species disappeared for the winter and was not observed again until water temperature increased in April.

Zhdanova also cultured this species in the lab and observed maximum growth rates during the first 5 to 10 days. *E. coregoni* molts once before reaching maturity and then once for each brood produced.

Additional information on the life history of this species follows:

embryonic development	2–3 days
sexual maturation	3–7 days
interval between egg deposition	2–3 days

no. of broods/female	8–9
life span	27–36 days
progeny/female	44
maximum molts	7
average length, newly hatched	0.26 mm
maximum length; female	0.54 mm
average wet weight; female	0.019 mg

Chydorus Leach 1843

TAXONOMIC HISTORY

The genus *Chydorus* contains at least five Great Lakes species. *C. sphaericus* is the most well known and has often been considered a common, easily identifiable species (Birge 1897). Recent studies (Frey 1980) present evidence that there are really several morphologically similar but distinct species in the *C. sphaericus* group, each with a restricted distribution. For convenience, in this study all of these recently recognized forms will be considered under the old name of *C. sphaericus* (O. F. Müller) 1785. The other chydorids found in the Great Lakes include *C. faviformis* Birge 1893, *C. globosus* Baird 1850, *C. latus* Sars 1862, and *C. gibbus* Sars 1891. *C. gibbus* has occasionally been recorded as *C. rugulosus* in the Great Lakes plankton literature (Forbes 1897; Frey 1982).

DESCRIPTION

All *Chydorus* are small cladocerans with spherical carapaces completely enclosing the body and limbs. The first antennae are covered by projections of the carapace (fornices) that unite with the rostrum to form a beak (Plate 20). The second antennae are biramous with three segments in each branch. The species are distinguished by the shape of the body and postabdomen, the pattern of the carapace, and the position of the olfactory setae on the first antennae. Variation in these characteristics have been observed in single populations of chydorids (Fryer 1968). Therefore, the detailed keys and descriptions by Scourfield and Harding (1941), Brooks (1959), Frey (1959), Smirnov (1966a), and especially Frey (1980) should be consulted when identifying chydorids to species.

SIZE

C. sphaericus is the smallest chydorid in the Great Lakes. We observed females from 0.3–0.4 mm long in Lake Superior (1979–1980). Brooks (1959) reported females ranging from 0.3–0.5 mm and males of 0.2 mm. The other species are slightly larger than *C. sphaericus* (Brooks 1959).

C. gibbus	♀ to 0.5 mm
C. faviformis	♀ 0.5–0.6 mm
C. latus	♀ 0.7–0.8 mm
C. globosus	♀ to 0.8 mm, ♂ to 0.6 mm

In Lake Michigan, *C. sphaericus* was found to have a dry weight of $0.8–1.2\mu g$ (Hawkins and Evans 1979).

DISTRIBUTION AND ABUNDANCE

Chydorids are distributed worldwide, occurring in waters of low salinity from near the equator to the Arctic Circle (Moore 1952). Each of the species in this genus has a more restricted distribution.

C. sphaericus has been found in all the Great Lakes (Table 11). This species is present in low numbers (< 1 m^{-3}) in Lake Superior (Selgeby 1975a and this study); in moderate numbers (8000–15000 m^{-3}) in Lakes Huron (Watson and Carpenter 1974), Ontario (McNaught and Buzzard 1973; Watson and Carpenter 1974), and Michigan (Gannon 1972a; Stewart 1974); and is occasionally quite abundant (3000–30000 m^{-3}) in Lake Erie (Bradshaw 1964; Rolan et al. 1973; and Heberger and Reynolds 1977). The abundance of this species in the Great Lakes may be underestimated because of the use of plankton nets with meshes 220μm or larger and the collection of samples from only the upper layers of the water column.

The other species of *Chydorus* are not as well known from the Great Lakes. *C. faviformis* has only been reported from Lake Huron (Bigelow 1922) while *C. latus* was collected once from Lake Erie (Wilson 1960). *C. globosus* occurs in low numbers in lakes Huron (Bigelow 1922) and Erie (Langlois 1954, Wilson 1960). Wilson (1960) found *C. gibbus* in Lake Erie while Forbes (1891), Birge (1893), and Selgeby (1974, 1975a) observed it in littoral samples from Lake Superior.

LIFE HISTORY IN THE GREAT LAKES

The life history and ecology of *C. sphaericus* is discussed in the following sections. The other four species occur in low numbers in the Great Lakes and have not been as well studied.

C. sphaericus is present year-round in Lake Erie but is generally only found in the summer and fall in the other Great Lakes (Watson and Carpenter 1974). Abundance peaks occur between August and November (Davis 1954, 1962; Bradshaw 1964; Rolan et al. 1973; Britt et al. 1973; Beeton and Barker 1974; Gannon 1974; Stewart 1974; Wat-

TABLE 11
Reports of *Chydorus sphaericus* in the Great Lakes

	Sampling Date	Abd[a]	Reference		Sampling Date	Abd[a]	Reference
LAKE ERIE	1882	U	Vorce 1882	**LAKE MICHIGAN (continued)**	1973–1974	U	Torke 1975
	1926	P	Langlois 1954		1974	C	Evans and Stewart 1977
	1928	PF	Sibley 1929		1975–1977	—	Hawkins and Evans 1979
	1928–1929	PF	Wilson 1960				
	1929	U–PF	Ewers 1933	**LAKE HURON**	1902, 1905, 1907	P	Bigelow 1922
	1948–1949, 1959	P	Bradshaw 1964		1967–1968	U	Patalas 1972
	1950–1951	C	Davis 1954		1970–1971	P	Carter 1972
	1956–1957	C	Davis 1962		1974–1975	P	McNaught et al. 1980
	1961	C	Britt et al. 1973				
	1967	P	Davis 1968	**LAKE ONTARIO**	1967	U	Patalas 1969
	1967–1968	C	Davis 1969		1967–1968	U	Patalas 1972
	1967–1968	P	Patalas 1972		1970	P	Watson and Carpenter 1974
	1968–1970	C	Heberger and Reynolds 1977		1972	P	McNaught and Buzzard 1973
	1971–1972	P–C	Rolan et al. 1973		1972	U–P	Czaika 1974a
LAKE MICHIGAN	1888	P	Eddy 1927		1973–1974	P	Czaika 1978a
	1926–1927	P	Eddy 1934	**LAKE SUPERIOR**	1889	U	Forbes 1891
	1966, 1968	P	Wells 1970		1893	P	Birge 1893
	1969–1970	P	Gannon 1972a		1928	R	Eddy 1934
	1969–1970	P	Gannon 1974		1971–1972	R–P	Selgeby 1975a
	1969–1970	P	Gannon 1975		1971	P	Selgeby 1974
	1971	P	Howmiller and Beeton 1971		1973	U	Upper Lakes Ref. Group 1977
	1971–1972	P	Beeton and Barker 1974		1979–1980	U	This study
	1972–1977	P	Evans et al. 1980				
	1973	P	Stewart 1974				

[a]Abundance Code
R = rare   U = uncommon   P = present   C = common   A = abundant   F = found in fish stomach contents
— = abundance ranking not appropriate

son and Carpenter 1974). During an October abundance peak in Lake Erie, the chydorid population was composed of 50% immature animals, and only 8% of the adult females were carrying eggs (Davis 1962).

ECOLOGY IN THE GREAT LAKES

Habitat. *C. sphaericus* is a nearshore and littoral species (Birge 1893; Wilson 1960; Czaika 1974a; Stewart 1974) found in lakes of all sizes. It can swim well for short distances but is usually found clinging to filamentous algae with its modified limbs (Birge 1897; Fryer 1968). Evans and Stewart (1977) believe that this is an epibenthic cladoceran, with the majority of the population moving up from the sediment surface only at night.

    C. sphaericus is much more abundant in Lake Michigan's southern Green Bay than in the northern half (Gannon

1974). This distribution pattern may be due to the influence of the Fox River, which empties into the southern part of the bay. Stewart (1974) agreed that the species was generally most abundant in such eutrophic areas.

Diurnal Migration. *C. sphaericus* is found in low numbers throughout the water column during the day. Abundance at all depths is 11 to 13 times greater at night. This suggests that the species migrates away from the bottom sediments mainly at night (Evans and Stewart 1977).

Feeding Ecology. Chydorids feed in two different manners. They can filter feed on small algal particles and detritus (Franke 1925; Smirnov 1962) or can attach to large algal filaments and macrophytes and scrape off attached detritus and larger diatoms (Smirnov 1962; Fryer 1968). The filtering and ingestion rates of *C. sphaericus* have been measured in Lake Huron (McNaught et al. 1980). This species consumes nannoplankton at an average rate of 2.1% of its body weight per day.

As Food for Fish.   *C. sphaericus* is eaten by largemouth bass, yellow perch, mosquitofish, bullheads, redhorse, and carpsuckers (Pearse 1921; Ewers 1933; Wilson 1960).

LIFE HISTORY AND ECOLOGY IN OTHER LAKES

Chydorids survive the winter months as adults and are capable of reproducing under the ice (Birge 1897; Smyly 1958). In Lake Mendota, Wisconsin, *C. sphaericus* showed no periodicity in its reproductive pattern. The population maxima were not related to physical factors such as light and temperature but did vary regularly with the relative abundance of *Anabaena* and other algae (Birge 1897).

This cladoceran is primarily a littoral (Moore 1952) and benthic form (Scourfield and Harding 1941; Fryer 1968; Husmann et al. 1978) that will migrate out of the substrate at night (Williams and Whiteside 1978). It has been found to number over 1400000 organisms m⁻³ on certain lake bottoms (Fryer 1968).

C. sphaericus is tolerant of a wide range of physical conditions. Its lethal level of dissolved oxygen is as low as 0.36 mg l⁻¹ at 19°C (Fryer 1968), and it has been found at pH levels from 3.4–9.5 (Ward 1940). Whiteside (1970) noted that *C. sphaericus* was generally most abundant in eutrophic lakes and least abundant in clear lakes, ponds, and bogs.

CALANOID COPEPODS

Senecella calanoides Juday 1923

TAXONOMIC HISTORY

Senecella calanoides is the only member of the family Pseudocalanidae found in fresh water; the other members of this family are marine. It is readily distinguished from other fresh-water calanoid copepods and has had a very stable taxonomic history.

DESCRIPTION

Adults of this large copepod are recognized by their short caudal rami with four stout terminal setae and one thin one (Plate 21). Unlike other calanoids, the mature males do not

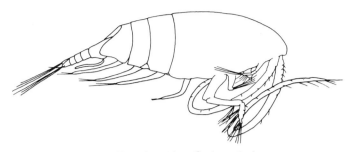

Figure 101   *Senecella calanoides*, ♀, lateral view

have geniculate right first antennae but do possess greatly enlarged and modified fifth legs (Plate 22). The females lack fifth legs and do not carry the eggs in egg sacs (Fig. 101). Mature females have urosomes of four segments and enlarged genital segments but are difficult to distinguish from immature CV female copepodids. Juday (1925) gives a more detailed description of the adults. The naupliar and copepodid stages of *S. calanoides* can generally be distinguished from those of other copepods by their relatively large size. Czaika (1982) provides keys to the identification of these stages.

SIZE

Marsh (1933) reports *S. calanoides* ranging from 2.4–2.6 mm long. In Lake Michigan, these animals can reach 2.9 mm (Torke 1975). In Lake Superior we found that males ranged from 2.3–2.8 mm and females from 2.6–3.3 mm.

DISTRIBUTION AND ABUNDANCE

S. calanoides occurs in fresh-water lakes of northern Asia and North America. It has been reported from Lakes Michigan, Huron, and Superior during the last thirty years (Table 12). Carter (1969) and Robertson (1966) consider that it is present in deep areas of all of the Great Lakes except Lake Erie. In most areas abundance is very low (< 1 m⁻³) (Wells 1970), but at times it may be one-fifth as abundant as *Limnocalanus macrurus* (Carter 1969).

LIFE HISTORY IN THE GREAT LAKES

This species has been collected year-round (Carter 1969; Selgeby 1975a; Torke 1975) but usually in such low numbers that it is difficult to determine changes in seasonal abundance (Wells 1960).

Carter (1969) studied the reproductive cycle of *S. calanoides* in Parry Sound of Lake Huron. Adults first appeared in late August. Breeding began immediately and extended through November when 15% of the females were still car-

TABLE 12
Reports of *Senecella calanoides* in the Great Lakes

	Sampling Date	Abd[a]	Reference		Sampling Date	Abd[a]	Reference
LAKE ERIE	No Reports			**LAKE ONTARIO**		RF	Pritchard 1931
LAKE MICHIGAN	1954–1955, 1958	U	Wells 1960	**LAKE SUPERIOR**	1968	P	Patalas 1972
	1954–1961	F	Wells and Beeton 1963		1969	P	Swain et al. 1970b
	1964	P	Robertson 1966		1970–1971	P	Conway et al. 1973
	1966, 1968	P	Wells 1970		1971	P	Selgeby 1974
	1971–1972	U	Beeton and Barker 1974		1971–1972	P	Selgeby 1975a
	1972–1977	P	Evans et al. 1980		1973	P	Upper Lakes Ref. Group 1977
	1973	R	Stewart 1974		1973	C	Watson and Wilson 1978
	1973–1974	P	Torke 1975		1974	U	Basch et al. 1980
LAKE HURON	1956	P	Robertson 1966		1979–1980	P–C	This study
	1967–1968	P	Carter 1969				

[a]Abundance Code
R = rare   U = uncommon   P = present   C = common   A = abundant   F = found in fish stomach contents

rying spermatophores. Because the eggs are shed directly into the water (Torke 1975), it is difficult to determine how many clutches of eggs are produced by each female. In Lake Michigan the first instar nauplii (NI) appear in November and grow to the first copepodid stage (CI) by March (Torke 1975). Naupliar growth is completed before the spring phytoplankton bloom occurs. The copepodids utilize the increasing food supply and reach the CIV stage by June and the adult CVI stage by August. The adults then mate and live through the winter.

In Lake Superior, the reproductive cycle is similar to that described above but is a month later, with adults first appearing in late September or October (Selgeby 1975a; Watson and Wilson 1978). This pattern of reproduction resembles that of *Limnocalanus macrurus*, except for a slight difference in the breeding period and the slower development rate of *S. calanoides*.

The male to female ratio is generally 1:1 in Lake Huron (Carter 1969), but in Lake Michigan females are more common from February to November (Torke 1975).

ECOLOGY IN THE GREAT LAKES

Habitat. *S. calanoides* is always found in deep, cold-water masses (Marsh 1933; Patalas 1971; Dadswell 1974). In Lake Superior this species prefers regions with a water depth greater than 30 meters. During the day the species remains well below the thermocline (generally below 60 m) with the younger copepodids located somewhat deeper in the water column than the adults (Carter 1969).

Diurnal Migration. This calanoid is negatively phototactic. It stays near the bottom during the day but begins movement toward the surface just before dark until ascent is blocked by the warm surface waters (Wells 1960). Conway et al. (1973) never found it above a depth of 30 m.

Feeding Ecology. The feeding ecology of this species is presently unknown in the Great Lakes.

As Food for Fish. *S. calanoides* has only been found in the stomachs of bloaters and ciscoes collected near the bottom of deep-water masses (Pritchard 1931; Wells and Beeton 1963).

LIFE HISTORY AND ECOLOGY IN OTHER LAKES

Dadswell (1974) studied this copepod's tolerance to several physical factors. It has an upper lethal temperature of 14.5°C and is absent from lakes with sharp oxygen stratifications. Its pH preference is narrow, 6–6.5. *S. calanoides* is a member of a marine family of copepods and exhibits a high upper lethal salinity limit of 17‰.

Limnocalanus macrurus Sars 1863

TAXONOMIC HISTORY

Limnocalanus macrurus auctus, which Forbes reported from Lake Superior in 1891, is synonymous with *Lim-*

nocalanus macrurus Sars 1863. *L. grimaldi* DeGuerne 1886 is simply a subspecies of *L. macrurus* (Bowman and Long 1968). Holmquist (1970) made a thorough study of the genus and found a wide range of variability in the characteristics that had been used to separate *L. grimaldi* from *L. macrurus*. She decided that both forms belonged to the same species, *Limnocalanus macrurus*, and that the variability in morphology was related to the amount of time inland populations had been separated from the sea.

DESCRIPTION

L. macrurus is a large calanoid copepod recognized by its elongate caudal rami with five terminal setae and by its enlarged maxillipeds. Mature females have a three-segmented urosome with an enlarged genital segment (Plate 23). Adult males have five-segmented urosomes and geniculate right first antennae (Plate 24). More detailed descriptions of the adults (Schacht 1898; Sars 1903; Gurney 1931; Marsh 1933) and copepodids and nauplii (Torke 1974; Czaika 1982) are available.

SIZE

In Lake Superior we found that the copepodids of *L. macrurus* could be separated into discrete size classes corresponding to their developmental stages. First instar copepodids are 1.2–1.3 mm long while adults (CVI) range from 2.4–2.9 mm. Wilson (1959) reported that in other lakes males (2.2–2.8 mm) were slightly smaller than females (2.2–3.2 mm). Hawkins and Evans (1979) reported dry weights of immature copepodids ranging from 1.7–13.5μg. Adult females (13.2–88.2μg) were usually heavier than males (16.7–55.8μg). Conway (1977) developed the following formula to calculate the dry weight of *L. macrurus* adults from their body length: $\log_{10}$ Dry Wt. (μg) $-$ 0.98L $-$ 0.79, where L=body length in mm.

DISTRIBUTION AND ABUNDANCE

L. macrurus occurs in both fresh and salt water. It is circumpolarly distributed and has been reported from Norway, Sweden, the Baltic and Caspian seas, and the Great Lakes of North America since the late 1800s. In recent years it has also been found in several lakes in Canada, Poland, Russia, and Arctic regions.

This large calanoid copepod has been found in all the Great Lakes (Table 13). It was once reported occurring in Lake Superior in large patches with over 100,000 organisms m^{-3} (Olson and Odlaug 1966), but generally averages 33 m^{-3} yearly (Selgeby 1975a) and is not considered the dominant zooplankton species in this lake (Watson and Wilson 1978). Population densities are moderate (600–4000 m^{-3}) in Lakes

Ontario (McNaught and Buzzard 1973; Watson and Carpenter 1974), Huron (Carter 1969), and the central region of Lake Michigan (Wells 1970). The species is nearly absent from the warmer waters of Lake Erie (Robertson 1966; Davis 1969) and from Green Bay and southern Lake Michigan (Gannon 1972a) where its survival may be affected by physical factors (temperature and oxygen content) or increased fish predation (Wells 1970; Gannon and Beeton 1971).

LIFE HISTORY IN THE GREAT LAKES

Females do not brood their eggs like most other copepods. The eggs are extruded directly into the water after fertilization and fall to the lake bottom where they stick to detritus particles. The first nauplius stage may be a poor swimmer and remains near the lake bottom where it is difficult to collect by normal zooplankton sampling methods.

The reproductive cycle of this species has been studied in all the Great Lakes. Mating, as detected by the presence of females with spermatophores, occurs after fall overturn (Torke 1975). The nauplii grow slowly during the winter on the limited supply of algae. During the spring phytoplankton bloom, they molt to the copepodid form and grow rapidly on the increased food supply (Gannon 1972a; Torke 1975). This allows growth from egg to adult in 6–8 months, with the adults generally living another 6 months.

The timing of the reproductive cycle is not uniform in all the Great Lakes, as shown in the accompanying table.

Period of Greatest Abundance

Lake	Adults with spermatophores	Nauplii	Mature Adults	Author
Huron	Oct.–Nov.	Nov.–Dec.	Mar.–June	Carter 1969
Michigan	Oct.–Mar.	Nov.–Mar.	Winter–Spring	Gannon 1972a; Torke 1975
Erie	Jan.–Mar.	—	Winter-Spring	Davis 1969
Ontario	As late as Mar.	—	Winter-Spring	Roff 1972
Superior	Jan.–Feb.	Mar.	Aug.–Nov.	Selgeby 1975a; Conway 1977

In the cold waters of Lake Superior, maturation and reproduction occur a few months later than in the other lakes. This copepod exhibits an extended breeding season in Lake Ontario, which may be due to the production of two clutches of eggs or to the late maturation of some of the adults (Roff 1972).

In Lakes Huron and Michigan the sex ratio is 1:1 (Carter 1969; Torke 1975), while in Lake Superior females outnumber males 2 to 1 (Conway 1977; this study).

TABLE 13
Reports of *Limnocalanus macrurus* in the Great Lakes

	Sampling Date	Abd[a]	Reference		Sampling Date	Abd[a]	Reference
LAKE ERIE	1919–1920	F	Clemens and Bigelow 1922	**LAKE HURON**	1919–1920	F	Clemens and Bigelow 1922
	1928	F	Sibley 1929		1956	P	Robertson 1966
	1929	F	Ewers 1933		1967	P	Patalas 1972
	1928–1930	P	Wright 1955		1967–1968	C–A	Carter 1969
	1928–1929	P	Wilson 1960		1970	P	Watson and Carpenter 1974
	1938–1939	P	Chandler 1940		1974	U	Basch et al. 1980
	1950–1951	P	Davis 1954		1974–1975	P	McNaught et al. 1980
	1951	P	Davis 1961				
	1957–1958	U–P	Gannon and Beeton 1971	**LAKE ONTARIO**	1919–1920	F	Clemens and Bigelow 1922
	1961	U–P	Britt et al. 1973		1927–1928	F	Pritchard 1929
	1968	U	Patalas 1972		1967	P	Patalas 1969
	1970	R	Watson and Carpenter 1974		1967	P	Patalas 1972
					1969–1970, 1972	P	McNaught and Buzzard 1973
LAKE MICHIGAN	1881	P	Forbes 1882		1970	P	Roff 1972
	1888	P	Eddy 1927		1970	C	Watson and Carpenter 1974
	1926–1927	P	Eddy 1934		1970–1971	P	Carter 1972
	1954–1955, 1958	R–C	Wells 1960		1971–1972	P	Wilson and Roff 1973
	1954	A	Wells 1970		1972	P	Czaika 1974a
	1954–1961	PF	Wells and Beeton 1963		1973	P	Czaika 1978a
	1964	P	Robertson 1966				
	1966	P	Wells 1970	**LAKE SUPERIOR**	1889	P	Forbes 1891
	1965	C	McNaught and Hasler 1966		1928	P	Eddy 1934
	1966–1967	F	Morsell and Norden 1968		1930–1934	A	Eddy 1943
	1968	C	Wells 1970		1964	C–A	Olson and Odlaug 1966
	1969–1970	P	Gannon 1972a		1968	P	Patalas 1972
	1969–1970	R	Gannon 1974		1969	A	Swain et al. 1970b
	1969–1970	P	Gannon 1975		1970–1971	A	Conway et al. 1973
	1971–1972	P	Beeton and Barker 1974		1971	A	Conway 1977
	1972–1977	C	Evans et al. 1980		1971–1972	C	Selgeby 1975a
	1973	C	Stewart 1974		1971–1972	P	Selgeby 1974
	1973–1974	C	Torke 1975		1973	C	Upper Lakes Ref. Group 1977
	1975–1977	—	Hawkins and Evans 1979		1973	C	Watson and Wilson
					1974	P	Basch et al. 1980
					1979–1980	C–A	This study

[a] Abundance Code
R = rare   U = uncommon   P = present   C = common   A = abundant   F = found in fish stomach contents
— = abundance ranking not appropriate

ECOLOGY IN THE GREAT LAKES

Habitat. *L. macrurus* is a cold-water stenotherm generally restricted to the hypolimnion of large, deep northern lakes (Wells 1960; Patalas 1969, 1972). Laboratory studies (Roff 1973) established an upper lethal temperature of 18°C, but in nature *L. macrurus* is seldom found in waters warmer than 14°C or with a dissolved oxygen content less than 5.1 mg/1 (Strøm 1946). Egg hatching is severely affected by low oxygen concentrations (Roff 1973). These limited tolerances to change in temperature and oxygen levels may make the species a useful indicator of eutrophication and water pollution (Wells 1970; Gannon and Beeton 1971).

Carter (1969) found that the older copepodid stages were located lower in the water column than the younger stages. As the surface water temperature increased during the summer, the entire population moved down, but the age class stratification remained the same.

Diurnal Migration. McNaught and Hasler (1966) studied the diurnal migration of *L. macrurus* in Lake Michigan. During isothermal conditions, ascent from the bottom began at midday, with the first animals reaching the surface at sunset. By midnight, 25% of the population was near the surface and remained there until sunrise when they began their descent. The migration pattern is similar during stratified conditions, but 75% of the population does not cross the thermocline. During early spring, when temperatures were below 4°C, vertical migration was not observed.

McNaught and Hasler (1966) hypothesized that the species has a visual pigment that is sensitive to the blue light region of the spectrum. Major migrations were noted when the logarithm of the blue light intensity was changing most rapidly. The occurrence of younger copepodid stages above the older ones in the water column suggests that light sensitivity may increase with age (Roff and Carter 1972; Conway 1977).

Feeding Ecology. *L. macrurus* is omnivorous, effectively filter feeding on particles from 4–24μm in diameter (Rigler 1972). The filtering rate increases as the animal grows (Kibby and Rigler 1973). In Lake Superior the diatoms *Melosira* and *Asterionella* and a chrysophyte, *Dinobryon*, comprise 80% of the stomach contents of this copepod (Berguson 1971). In Lake Huron it consumes nannoplankton at the low rate of 1.1% of its body weight per day (McNaught et al. 1980). Bowers and Warren (1977) recently observed the searching behavior of *L. macrurus* as it preyed upon a Lake Michigan copepod, *Diaptomus ashlandi*. The species uses its elongate maxillipeds to hold prey items, which include cladocerans, other copepods, and its own naupliar stages.

As Food for Fish and Other Organisms. *L. macrurus* has a very high organic matter and protein content with a low percentage of crude fiber and chitin (Birge and Juday 1922). This makes it a very nutritious food for fish. In the Great Lakes it has been consumed by alewife, cisco, emerald shiner, and whitefish (Schacht 1898; Clemens and Bigelow 1922; Pritchard 1929; Ewers 1933; Morsell and Norden 1968). In other lakes it is a major item in the diet of herring (Forbes 1883; Pritchard 1931; Wierzbicka 1953; Bityukov 1960) and the brook silverside (Forbes 1885). Intense predation pressure may have affected the abundance of this species in Lake Michigan (Wells 1970).

Adults are often observed with protozoans attached to their thorax and abdomen. Evans et al. (1979) identified these protozoans as *Tokophrya quadripartita*, a suctorian species.

LIFE HISTORY AND ECOLOGY IN OTHER LAKES

The reproductive behavior of *L. macrurus* was observed in the laboratory by Roff (1972). Shortly after mating, the female released 10–20 eggs, which settled to the bottom. The opaque, brown eggs became transparent just before hatching. Egg development was found to be temperature dependent, requiring 38 days to hatch at 0°C and only 17 days at 10°C. Temperatures greater than 16°C caused some inhibition in hatching. Cooley and Minns (1978) developed a formula to predict egg development time of several species of copepods at different temperatures. Their formula for *L. macrurus* was derived using data from Roff's studies. Roff also found that egg mortality was temperature dependent. Minimum mortality (2%) ocurred at 10°C, with mortality increasing to 78% at 16°C.

Growth rate depends on food supply (Comita 1956). When the food supply is low, the first nauplii to hatch consume large quantities of small phytoplankton. They grow rapidly and soon molt to the next stage, which allows them to consume larger items before they become food limited. The later-hatching nauplii encounter a very low biomass of algae of the appropriate size and grow much slower. This variation in growth rate can result in bimodal size distribution for the population.

In most lakes *L. macrurus* produces only one generation each year. In arctic Char Lake, Rigler (1972) found that most development from egg to adult took place under the ice. Six to seven months were required for this development, with the animals living a total of 10–16 months (Roff and Carter 1972).

Bioassays showed that this copepod is sensitive to high chlorine concentrations that are often found near the cooling water outflows of power plants (Latimer 1975). At temperatures of both 5 and 10°C the 30-minute TL50 was set at 1.54 mg chlorine/1.

Epischura lacustris S. A. Forbes 1882

TAXONOMIC HISTORY

Epischura lacustris was first collected by S. A. Forbes from Grand Traverse Bay in Lake Michigan in 1881. Although three other members of this genus occur in North America, *E. lacustris* is the only species found in the Great Lakes region.

DESCRIPTION

E. lacustris is recognized by its short, broad caudal rami terminating in three well-developed stout setae. These characteristics allow all of the copepodid stages to be distinguished from other fresh-water copepods. Adult females possess twisted three-segmented urosomes, while the males are distinguished by winglike projections from the second urosomal segment (Plate 25).

Forbes (1882) observed that *E. lacustris* was colorless in fall samples from Lake Michigan, although it appeared red in spring collections. In Lake Superior females have light orange ovaries and slightly darker colored genital segments during the September breeding season. More detailed descriptions of the adults (Schacht 1898), copepodids (Marsh 1899; Main 1962), and nauplii (Czaika 1982) are available.

SIZE

Wilson (1959) reported females from 1.8–2.0 mm and males from 1.4–1.6 mm. We measured all copepodid stages from samples collected from western Lake Superior and obtained the following lengths:

Stage	Range (mm)	Avg. (mm)
CI	0.56–0.76	0.67
CII	0.70–0.96	0.82
CIII	0.88–1.14	1.02
CIV	1.18–1.48	1.31
CV♀	1.40–1.74	1.60
CVI♀	1.72–1.88	1.79
CVI♂	1.36–1.68	1.53

The dry weight of copepodid stages CI–CV from Lake Michigan varied from 0.8–3.6 micrograms while adults (CVI) weighed 5.9–13.9 micrograms (Hawkins and Evans 1979).

DISTRIBUTION AND ABUNDANCE

E. lacustris is found from the North Central United States north to Winnipeg, Canada. Although found in all the Great Lakes (Table 14), this species is rarely collected from Lake Ontario and appears to be less abundant in Lakes Huron, Erie, and Michigan now than in the past. In 1950, samples from Lake Erie contained 4900 adults and 12900 young m⁻³ (Davis 1954). By the 1960s and 1970s this copepod was only observed occasionally and in much lower numbers (Davis 1968; Rolan et al. 1973; Britt et al. 1973). In Lake Superior it is absent from the midlake waters (Watson and Wilson 1978) but is found in moderate numbers in the Apostle Islands region (max. 107 m⁻³) (Selgeby 1974) and the western tip of the lake (max. 200 m⁻³) (this study). During September and October this species' distribution pattern extends furthest offshore.

LIFE HISTORY IN THE GREAT LAKES

E. lacustris breeds in the summer and fall (Gannon 1972a; Selgeby 1975a). After fertilization, the females shed the eggs, which settle to the bottom and overwinter. Nauplii hatch from the eggs from May to July (Davis 1954; Selgeby 1975a) and undergo rapid development. Adults mature by July, and most females are found carrying spermatophores from August to November (Gannon 1972a; Selgeby 1975a).

It is difficult to determine the exact time of egg production and the number of clutches produced since the females do not carry the eggs. In littoral areas without a cold hypolimnion, some of the eggs produced in July may hatch out and produce a second generation (Selgeby 1974; Torke 1975).

ECOLOGY IN THE GREAT LAKES

Habitat. *E. lacustris* is found in the littoral region of deep, clear lakes (Schacht 1898; Stewart 1974) and prefers the epilimnetic and metalimnetic waters during temperature stratification (Forbes 1891; Eddy 1943; Wilson 1960; Wells 1960; Conway et al. 1973). In August and September this calanoid is more widely distributed throughout the water column.

Diurnal Migration. *E. lacustris* is a strong swimmer and undergoes regular vertical migrations. It moves upward at night, becoming most concentrated at the surface 1–1.5 hours after sunset (Wells 1960).

Feeding Ecology. This species, like some other large calanoid copepods, is omnivorous, consuming both algae and small zooplankton (Torke 1975). In Lake Huron it is an important grazer on nannoplankton and filters between 0.059 and 0.526 ml·animal⁻¹·hour⁻¹ (McNaught et al. 1980).

As Food for Fish. When *E. lacustris* was abundant in Lake Erie, it was an important item in the diet of young fish including bass, crappie, ciscoes, sauger, freshwater drum, shiners, trout-perch, and yellow perch (Wickliff 1920, Clemens and Bigelow 1922; Sibley 1929; Ewers 1933). Small bloater also consume this species in Lake Michigan (Wells and Beeton 1963).

LIFE HISTORY AND ECOLOGY IN OTHER LAKES

E. lacustris is usually found in the upper waters of stratified lakes where food is most abundant (Marsh 1897). It prefers warm water and does not migrate down to the hypolimnion during the day, instead remaining in the metalimnion or epilimnion.

TABLE 14
Reports of *Epischura lacustris* in the Great Lakes

	Sampling Date	Abd[a]	Reference		Sampling Date	Abd[a]	Reference
LAKE ERIE	1919–1920	F	Wickliff 1920	**LAKE MICHIGAN (continued)**	1972–1977	P	Evans et al. 1980
	1919–1920	F	Clemens and Bigelow 1922		1973–1974	U	Torke 1975
	1928–1929	CF	Wilson 1960		1975–1977	—	Hawkins and Evans 1979
	1928	F	Sibley 1929				
	1928–1930	C	Wright 1955	**LAKE HURON**	1907	A	Sars 1915
	1929	F	Ewers 1933		1956	P	Robertson 1966
	1929	F	Kinney 1950		1967–1968	P	Carter 1969
	1938–1939	P	Chandler 1940		1968	P	Patalas 1972
	1950–1951	C–A	Davis 1954		1970–1971	P	Carter 1972
	1951	P	Davis 1961		1970	R	Watson and Carpenter 1974
	1956–1957	U	Davis 1962		1974	C	Basch et al. 1980
	1961	R	Britt et al. 1973		1974–1975	P	McNaught et al. 1980
	1967	R	Davis 1968				
	1968	P	Patalas 1972	**LAKE ONTARIO**	1919–1920	F	Clemens and Bigelow 1922
	1970	U	Watson and Carpenter 1974		1967	P	Patalas 1972
	1971–1972	P	Rolan et al. 1973				
LAKE MICHIGAN	1881	C	Forbes 1882	**LAKE SUPERIOR**	1889	A	Forbes 1891
	1887–1888	C	Eddy 1927		1928	P	Eddy 1934
	1926–1927	P	Eddy 1934		1930–1934	A	Eddy 1943
	1927	R	Eddy 1927		1967–1969	P	Swain et al. 1970b
	1954–1955	P	Wells 1960		1968	P	Patalas 1972
	1954–1961	P	Wells and Beeton 1963		1970–1971	C	Conway et al. 1973
	1964	P	Robertson 1966		1971	P	Selgeby 1974
	1966, 1968	R–C	Wells 1970		1971–1972	P	Selgeby 1975a
	1969–1970	P	Gannon 1974		1973	C	Upper Lakes Ref. Group 1977
	1969–1970	P	Gannon 1972a		1973	C	Watson and Wilson 1978
	1971–1972	P	Beeton and Barker 1974		1974	C	Basch et al. 1980
					1979–1980	C	This study

[a]Abundance Code
R = rare   U = uncommon   P = present   C = common   A = abundant   F = found in fish stomach contents
— = abundance ranking not appropriate

Marsh (1897) did not observe well-defined population peaks for this species in the summer or fall in Green Lake, Wisconsin. Eggs hatched in February and early March, with nauplii appearing until the end of March. The species then disappeared and was not collected again until June.

In Whitmore Lake, Michigan this species produces two generations a year (Main 1962). Resting eggs hatch in the spring with the copepodids maturing in early summer. They produce eggs that develop rapidly into mature adults, which reproduce in the fall.

The feeding behavior of *E. lacustris* was observed by Wong (1981) who studied animals collected from Pinks Lake, Quebec. He decided that the species consumes any prey item that it can detect, capture, and ingest. *Bosmina* were generally preferred over *Ceriodaphnia* due to their

ease of capture, although it was unable to handle *Bosmina* greater than 0.40 mm long.

Eurytemora affinis (Poppe) 1880

TAXONOMIC HISTORY

Eurytemora affinis was described by Poppe in 1880 from samples collected at river mouths in Germany. Since that time 14 other species have been added to this genus, ten of them found in North America. *E. affinis* is the only species found in the Great Lakes region, but *E. hirundoides* has commonly been reported from Canada and the eastern

United States. Marsh (1933) and De Lint (1922) proposed that these two species were distinct. Further taxonomic descriptions by Gurney (1933) led McClaren et al. (1969) and Katona (1971) to conclude that all reports of *E. hirundoides* in North America were actually *E. affinis*.

DESCRIPTION

The caudal rami of *E. affinis* are elongate and possess five terminal setae. This species is distinguished from *Limnocalanus macrurus* by its smaller size and shorter maxillipeds. Adult females have long, pointed metasomal wings and three-segmented urosomes while males have five-segmented urosomes, greatly enlarged fifth legs, and geniculate right first antennae (Plate 26).

Marsh (1933) and Wilson (1959) describe adults in more detail while Gurney (1933), Katona (1971), and Czaika (1982) provide details of the naupliar and copepodid stages.

SIZE

Mature females generally range from 1.1–1.5 mm long while the males are 1.0–1.5 mm (Wilson 1959). In Lake Superior we found that females (1.37 mm) averaged slightly larger than males (1.20 mm).

Gurney (1933) studied the development of a related species, *E. velox*, and found nauplii ranging from 0.1 mm (N1) to 0.4 mm (N6) and copepodids from 0.4 mm (CI) to 1.9 mm (CVI). Females were larger than males at all copepodid stages.

In Lake Michigan *E. affinis* copepodids (CI–CV) have a dry weight of 0.5–1.9 µg while adults weigh 3.9–5.3 µg (Hawkins and Evans 1979).

DISTRIBUTION AND ABUNDANCE

This calanoid is euryhaline, occurring in salt, brackish, and fresh water (Wilson and Tash 1966). It has been reported from Britain, Germany, the Volga River, and the Caspian, Baltic, and North seas (De Pauw 1973). In North America, it has been found in the Gulf of Mexico, along the Atlantic and Pacific coasts, and recently in the Great Lakes. Gurney (1931, 1933) suggests that the distribution of this species in Europe may be related to that of *Mysis* and *Limnocalanus*.

E. affinis has only been found in the Great Lakes since the 1960s (Table 15). It was first observed in Lakes Erie and Ontario and later in the upper lakes. The distribution pattern is rather irregular (Patalas 1969) with largest concentrations occurring in harbors and bays. It is the most abundant calanoid in the Duluth-Superior Harbor (Balcer,

unpublished) and the littoral zone of Green Bay (Gannon 1972) where numbers reach 1000 m^{-3}. In other areas abundance is generally much lower (McNaught and Buzzard 1973; Czaika 1974a; Watson and Carpenter 1974). This distribution pattern may result from the introduction of organisms into the bays and harbors when ocean-going ships pump out their brackish bilge water before loading cargo at Great Lakes ports. Large populations of *E. affinis* build up in these warmer, sheltered areas before spreading out into the open lake (Faber and Jermolajev 1966).

LIFE HISTORY IN THE GREAT LAKES

E. affinis exhibits a patchy horizontal distribution in the Great Lakes and is usually collected between April and January (Gannon 1972a; Wilson and Roff 1973; Beeton and Barker 1974; Czaika 1974a; Stewart 1974). Peaks of abundance have generally been observed during the summer and fall (Patalas 1969; Wells 1970; Wilson and Roff 1973; Rolan et al. 1973; Watson and Carpenter 1974) although a January peak was noted in Cleveland Harbor (Rolan et al. 1973).

E. affinis overwinters as eggs (Wells 1970) that begin hatching in April (Stewart 1974). The animals grow rapidly, maturing by June. Females with eggs and all copepodid stages are observed between July and September, making it difficult to determine the number of generations produced each year.

ECOLOGY IN THE GREAT LAKES

Habitat. *E. affinis* is a euryhaline species that prefers warm epilimnetic and littoral waters (Wells 1970; Stewart 1974). Evans and Stewart (1977) concluded that this was an epibenthic species, observing that 60% of the population entered the water column only at night.

Diurnal Migration. This calanoid moves upward in the water column at night. In October Wilson and Roff (1973) found significant positive correlations between vertical migration patterns of adults and copepodids and the incident radiation.

Feeding Ecology. In southern Lake Huron *E. affinis* filters nannoplankton at a rate of 0.299 ml·animal^{-1}·hour^{-1} (McNaught et al. 1980). Richman et al. (1980) found that this species can eat particles with a spherical diameter of 12–33 µm but prefers the larger cells 19–33 µm in diameter.

As Food for Fish. In Lakes Huron and Ontario *E. affinis* was found to comprise up to 90% of the diet of young smelt

TABLE 15
Reports of *Eurytemora affinis* in the Great Lakes

	Sampling Date	Abd[a]	Reference		Sampling Date	Abd[a]	Reference
LAKE ERIE	1961	U	Engel 1962a	**LAKE HURON**	1964	AF	Faber and Jermolajev 1966
	1961	P	Britt et al. 1973		1968	P	Patalas 1972
	1962	U	Faber and Jermolajev 1966		1967–1968	U–P	Carter 1969
	1967	U	Davis 1968		1970	R	Watson and Carpenter 1974
	1967–1968	U	Davis 1969				
	1968	P	Patalas 1972		1970–1971	U–P	Carter 1972
	1970	P	Watson and Carpenter 1974		1974–1975	U–P	McNaught et al. 1980
	1971–1973	P	Rolan et al. 1973	**LAKE ONTARIO**	1962	U	Faber and Jermolajev 1966
	1973–1974	P	Czaika 1978b		1964	C	Robertson 1966
LAKE MICHIGAN	1964–1965	P	Robertson 1966		1967	P	Patalas 1969
	1966, 1968	P	Wells 1970		1967	P	Patalas 1972
	1969–1970	P	Gannon 1974		1970	P	Watson and Carpenter 1974
	1969–1970	P	Gannon 1972a				
	1971	P	Howmiller and Beeton 1971		1971–1972	C	Wilson and Roff 1973
					1972	P	McNaught and Buzzard 1973
	1971–1972	P	Beeton and Barker 1974				
	1972–1977	P	Evans et al. 1980		1972	U–P	Czaika 1974a
	1973	P	Stewart 1974		1972–1973	R–P	Czaika 1978a
	1974	P	Evans and Stewart 1977	**LAKE SUPERIOR**	1968	P	Patalas 1972
	1975–1977	—	Hawkins and Evans 1979		1973	R	Upper Lakes Ref. Group 1977
					1979–1980	P	This study

[a]Abundance Code
R = rare   U = uncommon   P = present   C = common   A = abundant   F = found in fish stomach contents
— = abundance ranking not appropriate

(Faber and Jermolajev 1966). Low numbers of this copepod in the plankton suggest that the fish are actively selecting this species or are feeding on the greater concentrations located near the sediment/water interface.

LIFE HISTORY AND ECOLOGY IN OTHER LAKES

In Tisbury Great Pond (along the east coast of the United States) *E. affinis* produces two generations a year (Deevey 1948). Copepodids (CI) appear in November, stages II through V are present by January, and adults and a second generation of copepodids appear in March. All copepodid stages are present until June when only adults remain. The population disappears in July when water temperature reaches 24°C. The first generation develops in slightly over 2.5 months while development of the second generation is more rapid.

Egg development time is greatly affected by water temperature (McLaren et al. 1969):

Water Temperature (°C)	Development Time (days)	
	Range	Mean
0	10.79–10.84	10.81
3.26	6.35–6.55	6.47
7.39	3.52–3.83	3.71
11.83	2.27–2.38	2.32

A lethal temperature of 33°C has been established for this species by Bradley (1976).

Family Diaptomidae

TAXONOMIC HISTORY

During the late 18th and early 19th centuries these calanoid copepods were included in the genus *Cyclops*. *Diaptomus* was established as a genus by Westwood in 1836 and was brought into general use by Schacht (1897). Light (1938,

1939) suggested splitting the American diaptomids into several subgenera including *Leptodiaptomus* and *Skistodiaptomus*. In 1953 Kincaid proposed elevating diaptomid subgenera to genus level. The Great Lakes species names therefore changed as follows:

Diaptomus ashlandi Marsh 1893	= *Leptodiaptomus ashlandi*
Diaptomus minutus Lilljeborg 1889	= *Leptodiaptomus minutus*
Diaptomus sicilis S. A. Forbes 1882	= *Leptodiaptomus sicilis*
Diaptomus siciloides Lilljeborg 1889	= *Leptodiaptomus siciloides*
Diaptomus oregonensis Lilljeborg 1889	= *Skistodiaptomus oregonensis*
Diaptomus pallidus Herrick 1879	= *Skistodiaptomus pallidus*
Diaptomus reighardi Marsh 1895	= *Skistodiaptomus reighardi*

GENERAL INFORMATION ON DIAPTOMIDS

The caudal rami of diaptomids terminate in five caudal setae of equal length. Adult females have symmetrical fifth legs, while the right fifth leg of males is enlarged and terminates in a claw. Characteristics of the fifth legs and other appendages are used in species identification. Czaika and Robertson (1968) and Czaika (1982) provide keys to the copepodid stages of the Great Lakes diaptomids. The general structure of diaptomid nauplii is described by Comita and Tommerdahl (1960) and Comita and McNett (1976). Keys to Great Lakes nauplii are available in Czaika (1982).

The early nauplius stages of diaptomids live on stored yolk material and do not feed. The diets of the older copepodid stages of the diaptomid species are quite variable. Some collect only small algae and bacteria while others consume large algae, rotifers, or small crustacean zooplankton.

Water temperature affects the growth of diaptomids; animals developing in cold water are generally larger than those reared at warmer temperatures (Coker 1933). Adult body size also depends on the food supply, with maximum growth occurring when food is abundant (Siefkin and Armitage 1968).

Food is stored as highly colored lipid droplets during winter and spring when temperatures are cool. As the water begins to warm, metabolism changes and excess food is stored as carbohydrates. This metabolic change is responsible for the change in body color from bright red during the winter to light blue or colorless during the summer (Siefkin and Armitage 1968).

Most large lakes contain two or more diaptomid species. The animals generally differ in body size, have different food preferences, occupy different water strata, or occur at different times of the year (Hutchinson 1951; Sandercock 1967).

Leptodiaptomus ashlandi (Marsh) 1893

TAXONOMIC HISTORY

In the Great Lakes *Leptodiaptomus ashlandi* has usually been reported as *Diaptomus ashlandi*. Due to its resemblance to *Leptodiaptomus sicilis*, it has occasionally been referred to as *Diaptomus sicilis* var *imperfectus* (Forbes 1891).

DESCRIPTION

Adult females can easily be distinguished from other diaptomids by their two-segmented urosomes and asymmetrical, rounded metasomal wings. Males are characterized by a large lateral spine located at the proximal end of the terminal segment of the exopod of the right fifth leg. Their geniculate right first antenna has a slender projection from its antepenultimate segment (Plate 27). Detailed descriptions of adults are given by Marsh (1893) and Jahoda (1948), while Czaika and Robertson (1968) and Czaika (1982) describe all of the copepodid stages.

SIZE

In Lake Superior we found adult females averaging 0.99 mm long (range 0.9–1.1 mm) while males averaged 0.95 mm (range 0.9–1.0 mm). The dry weight of adult females in Lake Michigan varied from 2.1–7.1 µg while that of males ranged from 1.9–5.6 µg. Weight varied seasonally, with the lightest animals occurring during the summer (Hawkins and Evans 1979).

DISTRIBUTION AND ABUNDANCE

L. ashlandi is one of the most widely distributed North American diaptomid copepods, occurring from the east to west coasts of Canada and in the northern half of the United States.

This species is not evenly distributed throughout the Great Lakes (Table 16). It has been reported as the dominant calanoid copepod in open waters of Lake Michigan with numbers usually exceeding 1000 m^{-3} (Beeton and Barker 1974; Stewart 1974; Torke 1975). The only crustacean zooplankton species with a greater abundance in the open waters of Lake Michigan is *Diacyclops thomasi*. In in-

TABLE 16
Reports of *Leptodiaptomus ashlandi* in the Great Lakes

	Sampling Date	Abd[a]	Reference		Sampling Date	Abd[a]	Reference
LAKE	1928–1929	P	Wilson 1960	**LAKE**	1973–1974	A	Torke 1975
ERIE	1928	PF	Sibley 1929	**MICHIGAN**	1974	P	Evans and Stewart 1977
	1928–1930	P	Wright 1955	**(continued)**	1975–1976	P	Bowers 1977
	1929	PF	Ewers 1933		1975–1977	—	Hawkins and Evans 1979
	1938–1939	C	Chandler 1940				
	1946–1948	C	Jahoda 1948	**LAKE**	1956	P	Robertson 1966
	1950–1951	C	Davis 1954	**HURON**	1967–1968	A	Patalas 1972
	1951	C	Davis 1961		1967–1968	A	Carter 1969
	1961	P	Britt et al. 1973		1970–1971	P	Carter 1972
	1967–1968	P	Davis 1969		1974	P	Basch et al. 1980
	1968	P	Patalas 1972		1974–1975	P–C	McNaught et al. 1980
	1971–1972	P	Rolan et al. 1973				
	1973–1974	P	Czaika 1978b	**LAKE**	1972	P	McNaught and Buzzard
				ONTARIO			1973
LAKE	1926–1927	C	Eddy 1927		1972	U–P	Czaika 1974a
MICHIGAN	1926–1927	R	Eddy 1934		1972–1973	R	Czaika 1978a
	1954–1955	P	Wells 1960				
	1954–1961	F	Wells and Beeton 1963	**LAKE**	1913	F	Hankinson 1914
	1964	P	Robertson 1966	**SUPERIOR**	1928	R	Eddy 1934
	1966, 1968	P	Wells 1970		1967–1968	A	Patalas 1972
	1969–1970	P–C	Gannon 1974		1971	A	Selgeby 1974
	1969–1970	C–A	Gannon 1972a		1971–1972	C	Selgeby 1975a
	1969–1970	A	Gannon 1975		1973	P–C	Watson and Wilson 1978
	1971	P–C	Howmiller and Beeton		1973	C	Upper Lakes Ref. Group
			1971				1977
	1971–1972	C	Beeton and Barker 1973		1974	P	Basch et al. 1980
	1972–1977	C	Evans et al. 1980		1979–1980	P	This study
	1973	A	Stewart 1974				

[a] Abundance Code

R – rare   U = uncommon   P = present   C – common   A = abundant   F = found in fish stomach contents
— = abundance ranking not appropriate

shore waters of Lake Michigan, *L. ashlandi* is not as common, less than 500 m⁻³ (Stewart 1974), and it is scarce in Green Bay (Gannon 1972a).

In 1971 Selgeby (1974) found yearly averages of 1377 and 2211 *L. ashlandi* m⁻³ at two stations in western Lake Superior. Abundance decreased dramatically throughout the lake by 1973 (Watson and Wilson 1978), and by 1979–1980 we found that adult *L. ashlandi* always comprised less than 1% of the crustacean zooplankton community of Lake Superior. Maximum abundances of 60 m⁻³ were found in the western end of the lake.

The distribution of *L. ashlandi* also varies in Lake Erie. It was the second most abundant crustacean zooplankter in 1948 (Jahoda 1948), but adults only comprised 1% of the summer population in 1972 (Patalas 1972). This species is more common in the western basin of the lake and in the Cleveland area where peaks of 420 organisms m⁻³ were recorded (Rolan et al. 1973).

The species is extremely rare in Lake Ontario, with its first appearance in the shallow waters of the northeast sector of the lake occurring in 1972 (McNaught and Buzzard 1973).

LIFE HISTORY IN THE GREAT LAKES

Adult *L. ashlandi* are found year-round in the Great Lakes. They are most common during the spring and summer (Chandler 1940; Davis 1954; Wells 1960; Wilson 1960; Carter 1969, Gannon 1972a, 1975; Britt et al. 1973) although peaks have been observed as early as March (Rolan et al. 1973) and as late as October and November (Beeton and Barker 1974).

Although some animals are breeding at all times, the

basic reproductive pattern is bivoltine and accounts for the seasonal variations in abundance mentioned above. Females with eggs are most numerous in April or May (Davis 1961; Selgeby 1975a; Torke 1975). These eggs produce a large, slow-growing spring population coinciding with the phytoplankton maximum. Mortality is high, with a small proportion maturing and breeding in July and August. The resulting smaller second generation grows rapidly and matures by fall overturn. Mortality is low for these adults, which overwinter and begin breeding in the spring (Selgeby 1975a).

The sex ratio is generally 1:1 throughout the year (Jahoda 1948; Torke 1975) although males may occasionally be more common (Stewart 1974).

The eggs have a diameter of 10μm (Davis 1959a) with the number of eggs per clutch often increasing from April to June (Davis 1961). This is due to growth of the adults and a positive correlation between body size and clutch size.

ECOLOGY IN THE GREAT LAKES

Habitat. *L. ashlandi* is a pelagic copepod with a preference for clear and relatively cool water (Jahoda 1948). Although occurring in littoral areas, it is most abundant offshore in approximately 20m of water (Wilson 1960; Stewart 1974). *L. ashlandi* is often associated with *S. oregonensis*, as both occupy deep strata during the spring and fall and become metalimnetic during periods of stratification (Carter 1969; Stewart 1974).

Diurnal Migration. This calanoid species is generally absent from surface waters during the bright daylight hours but actively moves upward at night (Jahoda 1948; Wells 1960; Wilson 1960). While Jahoda found no seasonal differences in the migration pattern, Wilson observed the migration to be more intense in early summer than later in the year.

Feeding Ecology. *L. ashlandi* swims with its ventral surface upward as it filter feeds. Feeding trials showed that this species is size selective and prefers particles less than 10μm in size (Bowers 1977, 1980). Filtering and ingestion rates are proportional to temperature and chlorophyll concentrations during isothermal conditions of winter and early spring. During summer stratification, the ingestion rate is highest in the surface waters where the greatest concentration of particles less than 10μm are found. Richman et al. (1980) found that *L. ashlandi* feeds similarly to *L. siciloides* and consumes particles with an equivalent spherical diameter of 3–19μm. It does not consume *Stephanodiscus* cells of 30μm.

As Food for Fish. *L. ashlandi* has been found in small quantities in the stomachs of whitefish, perch, and shiners (Hankinson 1914; Sibley 1929; Ewers 1933). Wilson (1960) speculated that it might be a more important item in the diet of the small demersal fish that are seldom collected and examined. Further work by Wells and Beeton (1963) confirmed that *L. ashlandi* was an important food item of small, bottom-dwelling bloaters.

LIFE HISTORY AND ECOLOGY IN OTHER LAKES

L. ashlandi is a common diaptomid in Lake Washington, Washington, with abundance peaks reported in both the spring and fall (Scheffer and Robinson 1939). However, this species only produces one generation per year there (Comita and Anderson 1959). Major egg production occurs from January to March although some eggs are observed from October through August. It takes less than 1 year for growth from egg to adult. The nauplius stages are generally completed by mid-June, and the adult form is assumed by October. The adults overwinter and then reproduce in the spring.

The greatest concentrations of *L. ashlandi* in Lake Washington were found in the epilimnion during stratified periods. Phytoplankton concentrations were also highest in this region.

In Great Slave Lake *L. ashlandi* is believed to have a five-month life cycle with reproductive peaks in April and September (Moore 1979). Females from the summer population carry 25 eggs per clutch while those from the winter population only carry 10. Copepodid stages CIV to CVI were found to eat algae and detritus.

Leptodiaptomus minutus (Lilljeborg) 1889

TAXONOMIC HISTORY

This species has usually been referred to simply as *Diaptomus minutus*, but since the recent revisions in taxonomy it is now known as *Leptodiaptomus minutus*.

DESCRIPTION

Detailed descriptions of adults (Jahoda 1948; Wilson 1959), copepodid stages (Czaika and Robertson 1968), and nauplii (Czaika 1982) are available for this species. Adult females are small and have only two segments in the urosome. Their metasomal wings are symmetrical and rounded while the endopods of the fifth legs are greatly reduced in size. Adult males have geniculate right first antennae with a slender projection from the antepenultimate segment. There is a very small medial lateral spine on the exopod of the right fifth leg (Plate 28).

SIZE

L. minutus from Lake Superior are the same size as those reported by Wilson (1959). Adult females average 0.99 mm with a range of 0.9–1.1 mm while adult males average 0.91 mm with a range of 0.9–1.0 mm. In Lake Michigan adult females have dry weights of 1.7–4.9 μg while males weigh 1.8–4.4 μg. Lowest weights are found during the summer (Hawkins and Evans 1979).

DISTRIBUTION AND ABUNDANCE

This diaptomid is a northern species, occurring in Greenland, Newfoundland, Iceland, and the northeastern and midwestern United States and Canada (Marsh 1929; Schacht 1897).

It has been found regularly in all the Great Lakes (Table 17) but is most abundant in Lakes Huron (Patalas 1972) and Michigan (Gannon 1972a) where it ranks second only to *Leptodiaptomus ashlandi* during the winter months. Peaks of abundance may exceed a few thousand organisms per cubic meter in these lakes and Lake Erie. In 1971 Selgeby (1975a) reported a yearly average of 105 organisms m⁻³ at

TABLE 17
Reports of *Leptodiaptomus minutus* in the Great Lakes

	Sampling Date	Abd[a]	Reference		Sampling Date	Abd[a]	Reference
LAKE ERIE	1919–1920	PF	Wickliff 1920	**LAKE HURON**	1907	C	Sars 1915
	1928–1929	P	Wilson 1960		1956	P	Robertson 1966
	1928–1930	P	Wright 1955		1967–1968	P	Carter 1969
	1929	RF	Ewers 1933		1968	C	Patalas 1972
	1938–1939	P	Chandler 1940		1970–1971	C	Carter 1972
	1946–1948	C	Jahoda 1948		1970	P	Watson and Carpenter 1974
	1950–1951	C	Davis 1954		1974	C	Basch et al. 1980
	1951	P	Davis 1961		1974–1975	P–C	McNaught et al. 1980
	1956–1957	P	Davis 1962				
	1961	P	Britt et al. 1973	**LAKE ONTARIO**	1964	U	Robertson 1966
	1967–1968	P	Davis 1969		1967	P	Patalas 1969
	1968	P	Patalas 1972		1970	P	Watson and Carpenter 1974
	1970	P	Watson and Carpenter 1974		1971–1972	C	Wilson and Roff 1973
	1971–1972	P	Rolan et al. 1973		1972	P	McNaught and Buzzard 1973
	1973–1974	P	Czaika 1978b		1972	P	Czaika 1974a
LAKE MICHIGAN	1887–1888	C	Eddy 1927		1972–1973	C	Czaika 1978a
	1926–1927	C	Eddy 1927				
	1926–1927	C	Eddy 1934	**LAKE SUPERIOR**	1928	C	Eddy 1934
	1954–1961	UF	Wells and Beeton 1963		1968	P	Patalas 1972
	1954–1955, 1958	P	Wells 1960		1971–1972	P	Selgeby 1975a
	1964	P	Robertson 1966		1971	P	Selgeby 1974
	1966, 1968	C	Wells 1970		1973	P	Watson and Wilson 1978
	1969–1970	C	Gannon 1975		1973	P	Upper Lakes Ref. Group 1977
	1969–1970	P	Gannon 1974		1974	C	Basch et al. 1980
	1969–1970	P–C	Gannon 1972a		1979–1980	P	This study
	1971–1972	P	Beeton and Barker 1974				
	1972–1977	C	Evans et al. 1980				
	1973	P–C	Stewart 1974				
	1973–1974	C	Torke 1975				
	1975–1977	—	Hawkins and Evans 1979				

[a]Abundance Code
R = rare   U = uncommon   P = present   C = common   A = abundant   F = found in fish stomach contents
— = abundance ranking not appropriate

Sault Ste. Marie on Lake Superior. However, this species has become uncommon in Lake Superior (current study) and accounts for less than 1% of the plankton population between June and November. Numbers seldom exceeded 20 organisms m⁻³ during 1979 and 1980.

LIFE HISTORY IN THE GREAT LAKES

While occurring year-round in the lower four Great Lakes (Jahoda 1948; Wright 1955; Wilson and Roff 1973; Torke 1975), *L. minutus* is not found in Lake Superior during the winter months (Selgeby 1975a; this study). Seasonal abundance is variable with peaks reported in the summer (Stewart 1974; Watson and Carpenter 1974; Selgeby 1975a), fall (Watson and Carpenter 1974), and spring (Davis 1962; Rolan et al. 1973). Carter (1969) found three peaks in Lake Huron, while Jahoda (1948) found peaks occurring at nine-month intervals in Lake Erie. Recently Czaika (1974a) and Watson and Carpenter (1974) reported a uniform yearly distribution in Lake Ontario without obvious abundance peaks.

This variability in abundance is due to the differences in life histories in the Great Lakes. In Lake Superior (Selgeby 1975a) only one generation is produced each year. The animals hatch from resting eggs in May and June and experience high mortality rates before a few reach maturity in the fall and produce the overwintering resting eggs.

In Lake Michigan (Torke 1975) *L. minutus* is bivoltine with a cycle similar to *L. ashlandi* and *Skistodiaptomus oregonensis*. Adults overwinter and produce a slow-growing spring population coinciding with the spring phytoplankton bloom. This population reproduces in the summer and the rapidly growing second generation matures by fall overturn. In other areas (Davis 1961), reproduction appears continuous from spring through fall. The sex ratio for *L. minutus* is usually 1:1 (Jahoda 1948; Stewart 1974; Torke 1975). Unlike other diaptomid species, clutch size for this animal is negatively correlated with size of the females (Davis 1961).

ECOLOGY IN THE GREAT LAKES

Habitat. *L. minutus* is a cool-water species restricted to deep lakes at the southern end of its range (Marsh 1929). Distribution is often patchy (Beeton and Barker 1974; Torke 1975) with the animals more evenly distributed throughout the water column in the winter (Gannon 1975) and concentrating in the upper layers during summer stratification (Wells 1960; Stewart 1974). This pattern causes spatial overlap with *L. ashlandi* and *S. oregonensis*.

Diurnal Migration. Vertical migrations are common in this species but cannot be attributed strictly to the level of incident radiation (Jahoda 1948; Wells 1960; Wilson and Roff 1973). Several factors may be involved as the extent of the migration increases as the surface waters begin to warm in the spring (Wilson and Roff 1973).

Feeding Ecology. Like most diaptomids, this species is a filter feeder. Food particles are collected by the basketlike second maxillae (Fryer 1954). It feeds selectively, concentrating on the nannoplankton < 20μm in diameter. Due to its adaptations for feeding in high food concentrations, it is most commonly found grazing in eutrophic waters and areas of natural concentrations of phytoplankton (McNaught 1978). The filtering, ingestion, and assimilation rates of *L. minutus* in southern Lake Michigan have been determined by McNaught et al. (1980).

As Food for Fish and Other Organisms. This small diaptomid copepod is found occasionally in the stomachs of small bloater, crappie, yellow perch, and other fish (Wickliff 1920; Ewers 1933; Wilson 1960; Wells and Beeton 1963). It may be a more common prey for large zooplankton than for fish. Selgeby (1974) noted declines in the abundance of *L. minutus* when the predatory zooplankton, *Diacyclops thomasi*, *Mesocyclops edax*, *Epischura lacustris*, and *Leptodora kindti* were increasing in numbers in Lake Superior.

LIFE HISTORY AND ECOLOGY IN OTHER LAKES

In Newfoundland (Davis 1972) and Quebec (Boers and Carter 1978), *L. minutus* is bivoltine, but the timing of the generations is different from the pattern observed in Lake Michigan. The overwintering adults breed in early spring, producing a rapidly-growing population by July and August. This second generation develops more slowly and overwinters. Temperature and food are believed responsible for the differences in developmental rates of the instars of the two generations.

L. minutus is generally found in the upper waters of deep lakes (Marsh 1897; Patalas 1971). Although a slight trend toward regular diurnal migration was observed in Ontario lakes (Schindler and Noven 1971), a reverse migration pattern was observed under the ice (Cunningham 1972). A positive phototactic behavior may account for the single maximum observed at the surface during the day.

Studies of the developmental stages of *L. minutus* show that the first naupliar stage does not feed and that feeding is minimal in the second stage (Cooley 1978). More advanced instars select ultraplankton (<15μm) and nannoplankton (15–64μm) as food items (Boers and Carter 1978).

High mortality of the naupliar stages in Bluff Lake, Nova Scotia (Confer and Cooley 1977) has been attributed to predation by omnivorous zooplankton such as *Meso-cyclops edax*.

Cooley (1978) reported that female *L. minutus* in Bluff Lake are capable of producing two types of eggs. Subitaneous eggs are produced from early February until September. Females then produce resting eggs until October when egg production ceases for the year. The resting eggs require some time in cool water for hatching.

Leptodiaptomus sicilis (S. A. Forbes) 1882

TAXONOMIC HISTORY

Leptodiaptomus sicilis was previously known as *Diaptomus sicilis*. It is the American counterpart of the European species *Diaptomus gracilis* and is synonymous with the terms *D. pallidus* var *sicilis* Herrick, *D. natriophilus* Light and *D. tenuicaudatus* Marsh (Wilson 1959).

DESCRIPTION

In the spring *L. sicilis* is often distinctly colored, showing crimson bands on the body and blue-gray ovaries and eggs. Food reserves may be responsible for the coloration, as the animal is transparent at other times of the year (Forbes 1882).

Adult females are distinguished from other diaptomids by their three-segmented urosome and pointed, triangular metasomal wings. They lack expanded lateral projections on the genital segment. Males have a large medial lateral spine on the terminal segment of the exopod of the right fifth leg and a long slender projection from the antepenultimate segment of the right first antenna. Unlike *L. siciloides*, the males have pointed metasomal wings similar to the females (Plate 29).

More detailed descriptions are available in Wilson (1959), Czaika and Robertson (1968), and Czaika (1982).

SIZE

Forbes (1891) reported that female *L. sicilis* had a length of 1.3–1.6 mm. Wilson (1959) found a greater size range with females varying from 1.2–1.9 mm and males from 1.1–1.5 mm. In Lake Superior, we found females averaging 1.66 mm (range 1.5–1.8 mm) and males averaging 1.38 mm (range 1.3–1.4 mm).

Dry weight in Lake Michigan is 8.6–14.5 μg for adult males and 13.4–23.0 μg for females (Hawkins and Evans 1979).

DISTRIBUTION AND ABUNDANCE

This species is found in both fresh and saline waters from the east to the west coasts of the United States and Canada and from Alaska south to Missouri.

We found it to be the most abundant copepod in Lake Superior with abundance ranging from 20–3000 organisms m^{-3} in 1979–1980. Selgeby (1974, 1975a) reported yearly averages of 1688 and 1976 m^{-3} at two sites near the Apostle Islands and 803 m^{-3} at Sault Ste. Marie. In Lakes Ontario, Huron, and Erie *L. sicilis* is usually found in low numbers (Table 18). In Lake Michigan, abundance declined between 1954 and 1968 (Wells 1970). It now averages less than 100 m^{-3} in the open lake and is absent in Green Bay (Gannon 1972a).

LIFE HISTORY IN THE GREAT LAKES

L. sicilis is found during all seasons of the year (Eddy 1943; Davis 1954; Watson and Carpenter 1974; Beeton and Barker 1974; Selgeby 1974), but adults are generally most abundant between January and June (Chandler 1940; Jahoda 1948; Wilson and Roff 1973; Gannon 1975). In Lake Erie, abundance peaks have been recorded as late as July and August (Watson and Carpenter 1974), while in Lakes Michigan (Wells 1960) and Ontario (Wilson and Roff 1973) secondary peaks may occur in November.

The reproductive cycle is variable and may depend on water temperatures. In Lake Erie (Davis 1961) the breeding season lasts from October to April, but the greatest number of females with eggs are observed in January. The number of eggs per female is positively correlated with body size. In Lake Erie the male to female sex ratio is 1.72:1 (Jahoda 1948).

L. sicilis is probably univoltine in Lake Michigan (Gannon 1972a; Torke 1975). Although a few females with eggs are found in June and July, most reproduction occurs between November and May (Stewart 1974). The sex ratio may be 1:1 (Torke 1975) or may slightly favor females (Stewart 1974) in this lake.

The reproductive cycle of this species is best understood in Lake Superior (Selgeby 1975a). In this deep, cold water the animals mature in late summer or early fall and overwinter as adults. Although a few breed in the fall, the major reproductive activity doesn't begin until February or March. Reproduction peaks in the summer when 81–96% of the females are carrying eggs. Most females produce a clutch in May averaging 15 eggs. After these eggs hatch in June, a second clutch, averaging 24 eggs, is deposited in the egg sac. A few females survive until July and produce a third brood with fewer eggs. Some of the nauplii hatched in late May molt to the copepodid stage in June and grow

TABLE 18
Reports of *Leptodiaptomus sicilis* in the Great Lakes

	Sampling Date	Abd[a]	Reference		Sampling Date	Abd[a]	Reference
LAKE ERIE	1919–1920	F	Wickliff 1920	**LAKE MICHIGAN (continued)**	1973	P	Stewart 1974
	1919–1920	F	Clemens and Bigelow 1922		1973–1974	P	Torke 1975
	1928–1929	C	Wilson 1960		1975–1977	—	Hawkins and Evans 1979
	1928	F	Sibley 1929				
	1928–1930	P	Wright 1956	**LAKE HURON**	1920	F	Clemens and Bigelow 1922
	1929	P–F	Ewers 1933		1967–1968	U	Carter 1969
	1938–1939	A	Chandler 1940		1968	A	Patalas 1972
	1946–1948	C	Jahoda 1948		1970	P	Watson and Carpenter 1974
	1950–1951	A	Davis 1954				
	1951	C	Davis 1961		1970–1971	C	Carter 1972
	1956–1957	U	Davis 1962		1974	P	Basch et al. 1980
	1961	U	Britt et al. 1973		1974–1975	P–C	McNaught et al. 1980
	1967–1968	P	Davis 1969				
	1968	P	Patalas 1972	**LAKE ONTARIO**	1919–1920	F	Clemens and Bigelow 1922
	1970	C	Watson and Carpenter 1974		1967	U	Patalas 1969
					1967	U	Patalas 1972
	1971–1972	P	Rolan et al. 1973		1971–1972	C	Wilson and Roff 1973
	1973–1974	U	Czaika 1978b		1972	P	McNaught and Buzzard 1973
LAKE MICHIGAN	1881	A	Forbes 1882		1972	P	Czaika 1974a
	1882	P	Forbes 1888		1972–1973	P	Czaika 1978a
	1887–1888	P	Eddy 1927				
	1954–1955, 1958	C	Wells 1960	**LAKE SUPERIOR**	1889	A	Forbes 1891
	1954–1961	F	Wells and Beeton 1963		1930–1934	A	Eddy 1943
	1964	P	Robertson 1966		1968	A	Patalas 1972
	1966, 1968	P	Wells 1970		1971–1972	C–A	Selgeby 1975a
	1969–1970	P	Gannon 1972a		1971	A	Selgeby 1974
	1969–1970	P	Gannon 1974		1973	A	Watson and Wilson 1978
	1969–1970	P	Gannon 1975		1973	A	Upper Lakes Ref. Group 1977
	1971–1972	P	Beeton and Barker 1974		1974	C	Basch et al. 1980
	1972–1977	C	Evans et al. 1980		1979–1980	A	This study

[a] Abundance Code
R = rare   U = uncommon   P = present   C = common   A = abundant   F = found in fish stomach contents
— = abundance ranking not appropriate

rapidly to the CV stage; however the adult form is not reached until the end of August. The sex ratio in Lake Superior is generally close to 1:1, but females may be much more abundant in the spring when they are carrying eggs, and males may predominate later in the year (Selgeby 1975a; this study).

ECOLOGY IN THE GREAT LAKES

Habitat. *L. sicilis* is most commonly found in large, deep, clear and cold northern lakes (Jahoda 1948; Robertson 1966; Czaika 1974a; Stewart 1974). In Lake Ontario these animals are generally found in the cooler waters near the bottom of the water column (Wilson and Roff 1973).

This pattern results in complete spatial separation between this species and *L. minutus* (Wilson and Roff 1973).

L. sicilis is usually scarce in littoral zones (Carter 1969) but will move into shallow water in the winter months when the water mass has cooled to a uniform temperature (Johnson 1972; Czaika 1974a; Stewart 1974). Patalas (1972) believed this species decreased in abundance in Lake Erie as it became more eutrophic.

Diurnal Migration. *L. sicilis* is generally found deeper in the water column than other diaptomids and exhibits a better-defined pattern of vertical migration (Jahoda 1948). The animals migrate toward the surface at night, often traveling large distances before returning to their daytime depths. In July and August, the migrations are minimized as the animals remain lower in the water column throughout the day. By September vertical migrations seem to disappear. The change in migration pattern may be due to increasing light intensity or higher summer temperatures (Wilson and Roff 1973).

In Lake Superior a preliminary study (Eddy 1943) showed that occasionally *L. sicilis* was more abundant at the surface of the water mass than at greater depths.

Feeding Ecology. This diaptomid has the broadest food range of the Great Lakes species, ingesting particles from <10μm to >53μm in diameter (Bowers 1977). During winter and spring most phytoplankton is <10μm. At such times the ingestion rate of *L. sicilis* is proportional to the chlorophyll concentration. During the summer the >53μm size class of algae increases in the region of the thermocline. *L. sicilis* shows highest ingestion rates in this region while *L. ashlandi* feeds more effectively in areas with smaller food particles.

As Food for Fish. *L. sicilis* is a relatively important food for several species of fish including bloater, whitefish, ciscoes, bass, perch, crappie, shiners, brook silverside, trout-perch, and suckers (Forbes 1882; Wickliff 1920; Clemens and Bigelow 1922; Sibley 1929; Ewers 1933; Wilson 1960; Wells and Beeton 1963). In Lake Superior, juvenile smelt feed heavily on this species in inshore waters in the spring and early summer. However, Selgeby (1974) observed that the zooplankton population does not seem to be greatly reduced by this predation.

LIFE HISTORY AND ECOLOGY IN OTHER LAKES

L. sicilis is a widely distributed species. It is more common in larger lakes, such as Great Bear Lake, Canada where it generally composes 81% of the summer zooplankton community (Patalas 1971).

In Green Lake, Wisconsin, this species has a life history pattern similar to that of Great Lakes populations. Reproductive activity peaks between March and May (Marsh 1897), but immature copepodids persist in large numbers throughout the winter.

Leptodiaptomus siciloides (Lilljeborg) 1889

TAXONOMIC HISTORY

This species has generally been reported as *Diaptomus siciloides* in the Great Lakes zooplankton literature. It is now known as *Leptodiaptomus siciloides*. Wilson (1959) believes *Diaptomus cuahtemoci* Osterio Tafall (1941) may be synonymous with this species.

DESCRIPTION

Adult females of this species have prominent, pointed lateral wings on the genital segment of the three-segmented urosome (Plate 30). The large lateral spine on the exopod of the right fifth leg of the adult male is medial in position, and the antepenultimate segment of the right antenna has a short, thick projection. Males are distinguished from *L. sicilis* by lacking pointed metasomal wings (Plate 31). Detailed descriptions of the adults (Schacht 1897), copepodids (Comita and Tommerdahl 1960; Czaika and Robertson 1968), and nauplii (Ewers 1930; Czaika 1982) are available.

SIZE

Wilson (1959) found adult females from 1.0–1.3 mm long and males from 1.0–1.1 mm. We found CVI females from Lake Superior to be fairly uniform in length (1.2–1.3 mm) with an average of 1.25 mm. Adult males averaged slightly smaller (1.12 mm) with a range of 1.1–1.2 mm.

DISTRIBUTION AND ABUNDANCE

L. siciloides is found in lakes and ponds throughout North America except in the extreme north and along the east coast (Wilson 1959).

In Lake Erie this species is one of the most abundant calanoid copepods during the summer months, ranking second to *Skistodiaptomus oregonensis* (Davis 1961). Average abundance is approximately 1710 organisms m⁻³ (Davis 1968) with peaks of 15800 m⁻³ occurring near Cleveland (Rolan et al. 1973). *L. siciloides* was not found in Lake Michigan in 1964 (Robertson 1966) but had become fairly abundant (230–850 m⁻³) in Green Bay by 1969 (Howmiller and Beeton 1971; Gannon 1974). It is generally found in low numbers in Lakes Huron and Ontario (Table 19), although it may be abundant in sheltered bays (Tressler et al. 1953).

L. siciloides was collected from western Lake Superior by Watson (Upper Lakes Reference Group 1977). In our 1979–1980 sampling we also collected it in small numbers from the shallow waters near Duluth, Minn. and Superior, Wis.

LIFE HISTORY IN THE GREAT LAKES

L. siciloides has been found in the Great Lakes year-round (Jahoda 1948; Tressler et al. 1953; Gannon 1974) but is most abundant in the fall (Davis 1962; Rolan et al. 1973;

TABLE 19
Reports of *Leptodiaptomus siciloides* in the Great Lakes

	Sampling Date	Abd[a]	Reference		Sampling Date	Abd[a]	Reference
LAKE ERIE	1928–1930	P	Wright 1955	**LAKE HURON**	1967–1968	R	Patalas 1972
	1929	PF	Ewers 1933		1970	U	Watson and Carpenter 1974
	1938–1939	P	Chandler 1940				
	1946–1948	C	Jahoda 1948	**LAKE ONTARIO**	1939–1940	C	Tressler et al. 1953
	1950–1951	P	Davis 1954		1964	P	Robertson 1966
	1951	A	Davis 1961		1967	R	Patalas 1969
	1956–1957	P	Davis 1962		1967–1968	P	Patalas 1972
	1961	C	Britt et al. 1973		1972	P	Czaika 1974a
	1967	C	Davis 1968		1972–1973	R	Czaika 1978a
	1967–1968	C	Davis 1969				
	1967–1968	P	Patalas 1972	**LAKE SUPERIOR**	1973	R	Upper Lakes Ref. Group 1977
	1970	P	Watson and Carpenter 1974		1979–1980	R	This study
	1971–1972	A	Rolan et al. 1973				
	1973–1974	C	Czaika 1978b				
LAKE MICHIGAN	1969–1970	C	Gannon 1974				
	1971	C	Howmiller and Beeton 1971				

[a] Abundance Code

R = rare   U = uncommon   P = present   C = common   A = abundant   F = found in fish stomach contents

Watson and Carpenter 1974). Numbers decline in the winter and spring.

The reproductive cycle in the Great Lakes is not thoroughly understood. In Lake Erie breeding begins in May and an increasing number of females carry eggs from June through August (Davis 1961). Breeding activity then declines, and the last females with spermatophores are observed in December. The number of eggs per female is highest in June and July and has been positively correlated to body size of the female. Egg diameter remains fairly constant at 109 μm (Davis 1969). Males are most common from July to September (Davis 1964), when the sex ratio is 1:1 (Jahoda 1948).

Tressler et al. (1953) reported nauplii year-round but only in small numbers from January to March. In sheltered bays the nauplii seemed to prefer the epilimnion in spring and summer and moved to the hypolimnion in September. Distribution was more even throughout the water column during the rest of the year.

ECOLOGY IN THE GREAT LAKES

Habitat. *L. siciloides* is common in ponds and small lakes. It is thought that the species was originally washed into Lake Erie (Wright 1955) but has now become well established there (Davis 1961, 1968). It prefers warm, shallow, turbid waters (Jahoda 1948), occurring more abundantly in Green Bay than in the open waters of Lake Michigan (Howmiller and Beeton 1971). This distribution pattern may make it a potential indicator of eutrophication (Gannon 1972b).

Diurnal Migration. Migration patterns of this species are not distinct, with a tendency for the animals to occur in large patches rather than in definite vertical layers as other migrating species do. In turbid waters the animals appear to be more abundant at the surface during the day (Jahoda 1948).

Feeding Ecology. Comita and Tommerdahl (1960) report that *L. siciloides* eats food in the same size range that *Skistodiaptomus oregonensis* selects (100–800 μm³). Examination of the mouthparts suggests that this species is adapted for filter feeding and is not predatory. Further work by Richman et al. (1980) showed that *L. siciloides* can consume algae with an equivalent spherical diameter of 2.5–22 μm (100 to 6000 μm³) but prefers algae 12–22 μm in diameter. Filtering rates were also determined for the various algal size classes.

As Food for Fish. *L. siciloides* is consumed by largemouth bass, white bass, and yellow perch during the summer (Ewers 1933).

LIFE HISTORY AND ECOLOGY IN OTHER LAKES

In Severson Lake, Minnesota, Comita (1964, 1972) found that *L. siciloides* produces five generations in one year. Reproduction begins in April and ends in November. Development of the naupliar stages takes 5–15 days, while copepodid development lasts from 9–26 days. Development is most rapid in May. In June and July, females produce a single egg sac with 10–20 eggs seven days after they reach maturity. Adults only live 12–32 days and produce a single clutch of eggs. In laboratory cultures the animals were capable of producing eggs in only 4 days, and females could produce up to 8 broods in a lifetime.

L. siciloides has a broad tolerance for salinity and pH. It has been found in waters with salinities ranging from 625–25,911 ppm, the broadest range of all diaptomid species (Moore 1952). It can withstand pH values of 6.6–7.9 (Ward 1940) and has been found in a marsh in Louisiana where the pH reached 8.5 (Binford 1978).

This species is subject to invertebrate predation (Comita 1972). Large depressions in the population of *L. siciloides* correlated with population peaks of *Chaoborus* larvae.

Skistodiaptomus oregonensis (Lilljeborg) 1889

TAXONOMIC HISTORY

In the Great Lakes zooplankton literature, this species was consistently referred to as *Diaptomus oregonensis* until the subgenus name *Skistodiaptomus* was elevated to generic level.

DESCRIPTION

Adult females have three-segmented urosomes and lack metasomal wings. Adult males are characterized by a geniculate right antenna without a projection on the antepenultimate segment and a large subterminal lateral spine on the terminal segment of the exopod of the right fifth leg (Plate 32). More detailed descriptions of the adults are given by Jahoda (1948) and Wilson (1959), while Czaika and Robertson (1968) describe all copepod stages. Czaika (1982) has written a key to the nauplii of the Great Lakes copepods.

SIZE

In Lake Superior we found females ranging from 1.2–1.4 mm and averaging 1.33 mm. Males were somewhat smaller (range 1.1–1.3, avg. 1.19 mm). Wilson (1959) reported a size range of 1.2–1.5 mm for females and 1.2–1.4 mm for males.

In Lake Michigan CVI females have a dry weight of 3.8–10.9μg while males weigh 3.3 to 10.1μg (Hawkins and Evans 1979).

DISTRIBUTION AND ABUNDANCE

S. oregonensis is one of the most widely distributed diaptomid species, occurring throughout the northern United States and southern Ontario, Canada (Marsh 1893, 1929; Carl 1940).

It has been reported in all the Great Lakes (Table 20), and is the most common diaptomid in Lake Erie (Davis 1961, Wright 1955) where peaks in abundance occur in September. In 1968, this species averaged 700 organisms m^{-3} with a peak of 2700 m^{-3} (Davis 1968), and by 1973 abundance peaks had increased to 58000 m^{-3} (Rolan et al. 1973).

S. oregonensis was not collected from Lake Michigan in 1927 (Eddy 1927; Beeton 1965) but has now become the most common diaptomid in Green Bay (Gannon 1972a) and may often outnumber *Leptodiaptomus sicilis* in the open waters of the lake (Wells 1960). Peaks of 2580 organisms m^{-3} have been recorded in the summer in shallow areas of Lake Michigan (Howmiller and Beeton 1971).

This species has been found in Lakes Ontario and Superior only during the past 15 years. Average abundance in Lake Ontario is 30 m^{-3} (McNaught and Buzzard 1973), but it can become the dominant diaptomid species in the fall (Czaika 1974a). In Lake Superior adult *S. oregonensis* are uncommon, only appearing in fall samples and averaging <1m^{-3} (Selgeby 1975a, this study).

Overall, this copepod is abundant in the warmer, southerly portions of the Great Lakes and is of decreasing importance in the cooler, northerly regions (Robertson 1966).

LIFE HISTORY IN THE GREAT LAKES

S. oregonensis is present year-round in the Great Lakes (Jahoda 1948; Carter 1969; Wilson and Roff 1973; Gannon 1975). It is most abundant in the summer and fall but peaks have been observed from April to December. Gannon (1972a, 1974) reported abundance peaks as early as February in Green Bay.

This species produces two generations each year in Lake Michigan (Torke 1975). Adults overwinter and produce a slow-growing spring generation that coincides with the phytoplankton maximum. This generation reproduces in the summer, and the offspring grow rapidly, maturing by fall overturn. Most animals wait until spring to reproduce although Gannon (1972a) reported some winter reproduction in Green Bay. In Lake Superior, where abundance is

TABLE 20
Reports of *Skistodiaptomus oregonensis* in the Great Lakes

	Sampling Date	Abd[a]	Reference		Sampling Date	Abd[a]	Reference
LAKE ERIE	1928–1929	P	Wilson 1960	**LAKE MICHIGAN (continued)**	1973	C	Stewart 1974
	1928–1929	C	Wright 1955		1973–1974	P–C	Torke 1975
	1929	PF	Ewers 1933		1975–1977	—	Hawkins and Evans 1979
	1929	PF	Kinney 1950				
	1938–1939	C	Chandler 1940	**LAKE HURON**	1907	A	Sars 1915
	1946–1948	C	Jahoda 1948		1956	P	Robertson 1966
	1950–1951	C	Davis 1954		1967–1968	C	Carter 1969
	1951	A	Davis 1961		1968	P	Patalas 1972
	1956–1957	A	Davis 1962		1970	P	Watson and Carpenter 1974
	1961	P	Britt et al. 1973				
	1962	UF	Wolfert 1965		1970–1971	C	Carter 1972
	1967	C	Davis 1968		1974	C	Basch et al. 1980
	1967–1968	C	Davis 1969		1974–1975	P–C	McNaught et al. 1980
	1968	P	Patalas 1972				
	1970	P–C	Watson and Carpenter 1974	**LAKE ONTARIO**	1964	P	Robertson 1966
	1971–1972	A	Rolan et al. 1973		1967	P	Patalas 1969
	1973–1974	A	Czaika 1978b		1967	R	Patalas 1972
					1971–1972	P	Wilson and Roff 1973
LAKE MICHIGAN	1926–1927	A	Eddy 1927		1972	P	McNaught and Buzzard 1973
	1954–1955, 1958	F	Wells 1960		1972	P	Czaika 1974a
	1954–1961	F	Wells and Beeton 1963		1973	C	Czaika 1978a
	1964	P	Robertson 1966				
	1966, 1968	P	Wells 1970	**LAKE SUPERIOR**	1968	R	Patalas 1972
	1969–1970	C	Gannon 1972a		1971	U	Selgeby 1974
	1969–1970	P	Gannon 1974		1971–1972	R	Selgeby 1975a
	1969–1970	C	Gannon 1975		1973	P	Upper Lakes Ref. Group 1977
	1971	P	Howmiller and Beeton 1971		1973	P	Watson and Wilson 1978
	1971–1972	P	Beeton and Barker 1974		1974	C	Basch et al. 1980
	1972–1977	C	Evans et al. 1980		1979–1980	P	This study

[a] Abundance Code
R = rare   U = uncommon   P = present   C = common   A = abundant   F = found in fish stomach contents
— = abundance ranking not appropriate

low, *S. oregonensis* only produces one generation each year and overwinters as resting eggs (Selgeby 1974).

Sex ratios of 1:1 are most common (Jahoda 1948; Davis 1962, 1968; Torke 1975), but females may outnumber males (Stewart 1974). Clutch size appears to be positively correlated with size of the females (Davis 1961).

ECOLOGY IN THE GREAT LAKES

Habitat.   *S. oregonensis* is more abundant offshore than in the littoral zone (Stewart 1974). During stratified periods it concentrates above the metalimnion (Wells 1960; Carter 1969; Stewart 1974). In unstratified water masses and up-welling areas, distribution throughout the water column is more uniform (Wilson and Roff 1973). In Lake Erie (Wilson 1960) *S. oregonensis* has been found in swarms in the littoral area. Overall, this species has considerable habitat overlap with *Leptodiaptomus minutus* and *L. ashlandi* (Rigler and Langford 1967; Carter 1969).

Diurnal Migration.   This diaptomid generally migrates toward the surface at night (Wells 1960), but the migration pattern is positively correlated with incident radiation levels only in the fall (Wilson and Roff 1973). Daytime and early evening ascents have been observed under conditions of low transparency (Jahoda 1948).

Food and Feeding Behavior. After *S. oregonensis* molts to the first copepodid stage, it can filter feed on the same size class of particles $(100-800\mu m^3)$ as the adults (McQueen 1970; Comita and McNett 1976). This same size class is eaten by *Leptodiaptomus siciloides*.

Richman (1964, 1966) has determined the filtering and energy transformation rates of this species under various conditions using C^{14}-labelled food particles. In 1980 Richman et al. found that adult females from Green Bay can consume algae from $2.5-300\mu m$ in diameter $(100-1400\mu m^3)$. They preferred large cells $12-24\mu m$ in diameter and could consume large diatoms such as *Stephanodiscus* that were too large for *L. siciloides*.

As Food for Fish. *S. oregonensis* has been found in the stomach of several species of fish including bass, crappie, carp, suckers, freshwater drum, trout-perch, yellow perch, and whitefish (Ewers 1933; Wilson 1960). While comprising only a small percentage of the diet of some fish, it was found to be an important food item for small bloater (Wells and Beeton 1963).

LIFE HISTORY AND ECOLOGY IN OTHER LAKES

This diaptomid is generally absent from the deep, transparent lakes in the Experimental Lakes Area of Ontario (Patalas 1971), which are ecologically similar to the Great Lakes.

The reproductive pattern varies both within and between areas. Yearly changes in the pattern were observed in Sunfish Lake, Ontario (Lai and Carter 1970). In some years this species produced one generation and overwintered as a resting stage while in others development was slowed in winter but no diapause occurred. In some years two or three generations were produced. In Lake Mendota, Wisconsin this species overwinters as an adult to produce a single generation in the spring (Birge 1897). Two generations are produced each year in Lake Cromwell, Quebec (Paquette and Pinel-Alloul 1982) while three generations are seen in meromictic Teapot Lake, Ontario (Rigler and Cooley 1974).

The reproductive pattern depends on the food supply available. A prolonged food shortage has been found to trigger the production of resting eggs rather than the regular subitaneous eggs (Carter 1972). Food shortages may be due to competition from cladocerans during their summer blooms. Once resting eggs are formed, they require a cold period of 4°C before they can hatch (Cooley 1971).

CYCLOPOID COPEPODS

Acanthocyclops vernalis (Fischer) 1853

TAXONOMIC HISTORY

Acanthocyclops vernalis is one of the most common, most variable, and most frequently misidentified species of cyclopoid copepod in North America (Yeatman 1944). It has been listed as *Cyclops vernalis*, *C. americanus*, and *C. brevispinosus* in many Great Lakes zooplankton studies. Yeatman (1944) examined museum collections and waters containing these species from all over North America. In some cases, he found populations with all three forms as well as intermediate forms present in the same lake. Price (1958) showed that such morphological variation in individual populations may be due to development under different environmental conditions. Later researchers collectively referred to all of these forms as *C. vernalis*. The genus *Cyclops* is now considered to be composed of several genera (Kiefer 1960). Kiefer's (1929b) subgenus *Acanthocyclops* has been elevated to genus level, resulting in the classification of this animal as *A. vernalis*.

In the early 1900s *Cyclops viridis*, a European species, was also reported from Lake Superior (Hankinson 1914, Eddy 1934). This species is seldom found in the United States, suggesting that the authors simply misidentified *Acanthocyclops vernalis*.

A recent report by Kiefer (1978a) infers that the Great Lakes animals may actually be another species, *Acanthocyclops robustus*. He found that *A. vernalis* is strictly a littoral species while *A. robustus* may also be found in pelagic zones. He describes *A. robustus* as possessing a slimmer genital segment with less-developed metasomal wings than in *A. vernalis*. Since further work is needed on the taxonomy of the Great Lakes animals, we refer to them here as *A. vernalis*.

DESCRIPTION

A. vernalis has four terminal setae on each caudal ramus, with only the medial pair of each ramus elongated. The lateral seta is located near the posterior one third of the caudal ramus, which lacks hairs on its inner margin (Plate 33). This species is more robust than *Diacyclops thomasi* and has less-pronounced metasomal wings. The fifth legs of these two species are quite distinct (Yeatman 1959) and may be examined to confirm the identity of the animals.

More detailed descriptions of *A. vernalis* are given by

TABLE 21
Reports of *Acanthocyclops vernalis* in the Great Lakes

	Sampling Date	Abd[a]	Reference		Sampling Date	Abd[a]	Reference
LAKE ERIE	1928–1929	C	Wilson 1960	**LAKE HURON**	1907	P	Sars 1915
	1928–1930	P	Wright 1955		1967–1968	P	Carter 1969
	1929	F	Ewers 1933		1968	U	Patalas 1972
	1929	F	Kinney 1950		1970	U	Watson and Carpenter 1974
	1938–1939	P	Chandler 1940		1970–1971	P	Carter 1972
	1946–1947	C	Andrews 1953		1974–1975	P	McNaught et al. 1980
	1950–1951	C	Davis 1954				
	1961	C	Britt et al. 1973	**LAKE ONTARIO**	1967	P–C	Patalas 1969
	1962	F	Wolfert 1965		1967	U	Patalas 1972
	1967	C	Davis 1968		1969, 1972	P	McNaught and Buzzard 1973
	1967–1968	U	Davis 1969		1970	P	Watson and Carpenter 1974
	1968	P	Patalas 1972		1972	U	Czaika 1974a
	1968, 1970	C	Heberger and Reynolds 1971		1972–1973	U	Czaika 1978a
	1970	C	Watson and Carpenter 1974				
	1971–1972	C	Rolan et al. 1973	**LAKE SUPERIOR**	1913	F	Hankinson 1914
	1973–1974	P	Czaika 1978b		1928	P	Eddy 1934
					1968	U	Patalas 1972
LAKE MICHIGAN	1926–1927	P	Eddy 1934		1971	P	Selgeby 1974
	1966, 1968	P	Wells 1970		1971–1972	P	Selgeby 1975a
	1969–1970	P	Gannon 1972a		1973	P	Watson and Wilson 1978
	1969–1970	P	Gannon 1974		1973	P–C	Upper Lakes Ref. Group 1977
	1969–1970	P	Gannon 1975		1979–1980	P–C	This study
	1971	C	Howmiller and Beeton 1971				
	1972–1977	P	Evans et al. 1980				
	1973	P	Stewart 1974				
	1974	C	Evans and Stewart 1977				
	1975–1977	—	Hawkins and Evans 1979				

[a]Abundance Code
R = rare   U = uncommon   P = present   C = common   A = abundant   F = found in fish stomach contents
— = abundance ranking not appropriate

Yeatman (1944) and Kiefer (1978). Czaika (1982) has prepared a key to the immature copepodids and nauplii.

SIZE

In Lake Superior we found adult females from 1.0–1.4 mm long and males from 0.8–1.0 mm. Spring samples were 0.1 and 0.2 mm longer than summer samples.

In Lake Michigan Hawkins and Evans (1979) found adult males with a dry weight of 2.4–2.6μg. Adult females were heavier, with dry weights of 4.8–6.4μg.

DISTRIBUTION AND ABUNDANCE

A. vernalis is widely distributed throughout North America, occurring in water bodies of all sizes.

Although *A. vernalis* is found in all the Great Lakes (Table 21), it is generally found in low numbers (8–73 m[-3]) (Carter 1969; Wells 1970; McNaught and Buzzard 1973) and makes up less than 1% of the crustacean zooplankton community (Patalas 1972). Greater densities (600–18000 m[-3]) are found in more eutrophic areas with a high rate of resource turnover, such as western Lake Erie, Cleveland Harbor, and Lake Michigan's Green Bay (Davis 1968; Howmiller and Beeton 1971; Rolan et al. 1973).

LIFE HISTORY IN THE GREAT LAKES

A. vernalis is absent or present in very low numbers in zooplankton samples collected between December and May (Chandler 1940, Andrews 1953; Rolan et al. 1973; Selgeby 1975a). Abundance increases rapidly during the warmer months, with peaks generally observed between June and August (Andrews 1953; Davis 1954; Patalas 1969; Britt et al. 1973; Gannon 1974; Watson and Carpenter 1974; Heberger and Reynolds 1977).

The life cycle of this cyclopoid copepod is not well understood in the Great Lakes due to the difficulty of distinguishing the copepodids from those of the more abundant *Diacyclops thomasi*. In Lake Erie, adults appear when the water temperature reaches 8–10°C, and reproduction is observed until November when the temperature again cools to 10°C (Ewers 1936; Andrews 1953). The greatest abundance is generally found at temperatures near 20°C.

In Lake Superior, Selgeby (1974, 1975a) concluded that only one generation is produced each year. The animals may overwinter as diapausing CIV or CV copepodids, which break diapause when the temperature warms during the spring, then molt to the adult CVI stage during the summer. Females with large clutches of eggs are present between June and August. However, no new adults appear before winter, suggesting that it is the copepodid stages that overwinter.

Andrews (1953) found that female *A. vernalis* in Lake Erie outnumbered males 9.5 to 1 during most of the year, while Davis (1968) found slightly more males in July.

ECOLOGY IN THE GREAT LAKES

Habitat.   *A. vernalis* is most abundant in warm, eutrophic nearshore areas and in shallow harbors and bays (Howmiller and Beeton 1971; Wilson 1960; Gannon 1972a, 1974, 1975; Stewart 1974, Heberger and Reynolds 1977). It is only found in open water in the western basin of Lake Erie (Davis 1968, 1969; Patalas 1972). This species is often found in larger numbers in the bottom mud than in the overlying water column, suggesting that it is an epibenthic organism (Sars 1915; Wilson 1960; Evans and Stewart 1977).

Diurnal Migration.   The vertical distribution patterns of this species have not been well studied. Evans and Stewart (1977) noted that *A. vernalis* was 11–13 times more abundant in the plankton at night than during the day, suggesting that it migrates up from the sediment at night.

Feeding Ecology.   McNaught et al. (1980) present data on the filtering, ingestion, and assimilation rates of *A. vernalis* feeding on nannoplankton and net plankton in Lake Huron.

As Food for Fish.   *A. vernalis* is a moderately important food item for many species of fish including bass, crappie, sauger, freshwater drum, trout-perch, yellow perch, and young whitefish (Hankinson 1914; Ewers 1933; Kinney 1950). Its preference for the bottom sediments makes

A. vernalis available to bottom-feeding fish such as suckers (Wilson 1960).

LIFE HISTORY AND ECOLOGY IN OTHER LAKES

A. vernalis is usually found in the middle of the water column in shallow lakes (Marsh 1895; Patalas 1971). It is capable of reproducing under the ice (Marsh 1895; Coker 1934a, 1934b, 1934c; Ewers 1936), with 50% of the females of some lakes carrying eggs in January (Marsh 1895).

This species is adapted to rapid reproduction, allowing it to take advantage of a sudden increase in the food supply. Mature males are short-lived but produce many spermatophores (Price 1958). Laboratory studies (Ewers 1936; Price 1958) showed that females deposit 40–80 fertile eggs into the egg sacs within 5 days of mating. The eggs hatch in 20–30 hours and the female can produce a second brood within a day. Some females were observed to produce new broods every 1.5 days for up to 4 weeks.

Egg development rate is temperature dependent (Coker 1934b; Ewers 1936; Aycock 1942). At warm temperatures (20°C) eggs developed into adults in 7–8 days with a 50% mortality rate. These animals were fairly small, approximately 1.0 mm long. At cooler temperature (7–10°C) development took longer (44 days), and mortality was much higher (92%) but the surviving adults were larger, up to 1.5 mm. Robertson et al. (1974) and Hunt and Robertson (1977) cultured *A. vernalis* in the laboratory at several different temperatures. They report the % hatch, egg development time, time to maturation, mean clutch size, and total egg production at each temperature. Their development times were longer than those previously reported, possibly due to differences in the diet fed to the copepods.

Coker (1934b) found that this species became dormant at temperatures of 27–36°C, and came out of dormancy when the temperature decreased. Many died at 36°C, with males being less tolerant of high temperatures than females.

Feeding experiments showed that *A. vernalis* copepodid stages CIV and CV readily prey on the nauplii and copepodids of *Leptodiaptomus tyrrelli* and *L. sicilis* (Anderson 1970). The smallest items were consumed first. It seldom consumed *Daphnia pulex* or other cyclopoids. The feeding rate of this species was twice that of *Diacyclops thomasi*.

Later studies found that adult *A. vernalis* will consume their own nauplii and cladocerans (Robertson et al. 1974). Brandl and Fernando (1974) give a detailed description of the capture and ingestion of *Ceriodaphnia reticulata*, while Kerfoot (1978) describes the capture and handling of *Bosmina* by this cyclopoid.

Diacyclops thomasi (S. A. Forbes) 1882

TAXONOMIC HISTORY

In 1882 S. A. Forbes described a cyclopoid copepod that a Mr. Thomas had collected from Lake Michigan as a new species, *Cyclops thomasi*. Later work by E. B. Forbes (1897) listed the Great Lakes specimens as *Cyclops bicuspidatus*, a well-known European species. Gurney (1933) concluded that these animals belonged to the same species, *C. bicuspidatus*, but that the form commonly found in America was distinct enough to deserve status as a subspecies, *C. bicuspidatus thomasi*. Yeatman (1959) agreed with this classification. Most recent researchers have used Yeatman's key to the cyclopoids and referred to the Great Lakes specimens as *C. bicuspidatus thomasi*, although a few continued to call them simply *C. bicuspidatus*.

This species belonged to the subgenus *Diacyclops* as described by Kiefer (1929b), who later (1960) elevated it to genus level. Reed (1963), Dussart (1969), and Kiefer (1978a) concluded that the lake-dwelling *Diacyclops bicuspidatus thomasi* was morphologically and ecologically distinct from *D. bicuspidatus*, a temporary pond species, and gave it status as a full species, *D. thomasi*. We follow the latest classification and refer to the Great Lakes animals as *Diacyclops thomasi*.

DESCRIPTION

This species is similar to *Acanthocyclops vernalis*. Both possess four terminal setae on each caudal ramus with the two median setae of each ramus elongate. *Diacyclops thomasi* is distinguished from *A. vernalis* by having the lateral seta located near the midpoint of the caudal ramus instead of near the posterior fourth of the ramus (Plate 34). Yeatman (1959) also shows the differences in structure of the fifth legs of these two species. More detailed descriptions of adults are given by Forbes (1882) and Kiefer (1929b, 1978b). Czaika (1982) provides keys to identify the copepodids and nauplii of this species.

SIZE

Yeatman (1959) examined the size range of this species, finding adult females from 0.9–1.2 mm and males approximately 0.8 mm long. In Lake Superior we found a wider range of sizes, adult females measuring 1.0–1.4 mm and males 0.8–1.1 mm.

In Lake Michigan, CVI males had a dry weight of 1.2–2.9μg while females were heavier, 1.9–5.6μg (Hawkins and Evans 1979).

DISTRIBUTION AND ABUNDANCE

D. thomasi is one of the most common and widely distributed copepods in North America. It is a cold-water species, found in winter and spring in the southern parts of its range and year-round only in deep northern lakes.

On a yearly basis, *D. thomasi* is the most important species of crustacean zooplankton in the Great Lakes (Patalas 1972; Gannon 1972a; Stewart 1974). High densities have been reported from all of the Great Lakes (Table 22). Lake Erie has 300–30000 organisms m^{-3} (Chandler 1940; Andrews 1953; Davis 1954; Wells 1970; Rolan et al. 1973; Watson and Carpenter 1974; Stewart 1974; Torke 1975; Heberger and Reynolds 1977; Czaika 1978a, 1978b). A distinct decline in numbers occurs in the late summer and fall (Andrews 1953, Wright 1955; Wilson 1960; Wells 1960; Swain et al. 1970b; Rolan et al. 1973). A few studies have found abundance peaks in the late summer (Gannon 1972a; Carter 1969; Wilson and Roff 1973; Beeton 1965). Seasonal temperature variations may account for the delayed occurrence of population peaks during some years (Carter 1969).

LIFE HISTORY IN THE GREAT LAKES

All six copepodid instars, including reproductive adults, have been collected throughout the year (Gannon 1972a; Stewart 1974; Torke 1975; Selgeby 1975a). Selgeby (1974, 1975a) studied the life cycle of *D. thomasi* in detail in Lake Superior. He found two generations each year. Copepodid stages CIII to CV overwinter in an active form with some molting to the adult (CVI) stage by February. These animals mate by May or June and produce at least two clutches of eggs. The summer generation that develops from these eggs grows rapidly and matures during July and August. Resulting females produce fewer, though larger, eggs than the previous generation, and these eggs develop into the overwintering copepodids.

Torke (1975) found that *D. thomasi* also winters as active copepodids in Lake Michigan. Reproduction begins 3 months before thermal stratification occurs in the spring, allowing the production of three generations each year. Torke proposed that his data could also be interpreted as showing continuous reproduction between July and November.

Both Selgeby (1974, 1975a) and Torke (1975) observed highest mortality in the summer populations. This mortality may be due to fish predation.

Males may have a shorter lifespan than females (Torke 1975). Females are generally more abundant (Stewart 1974), and the female to male ratio often exceeds 2:1 (Andrews 1953; Davis 1962, 1968; Torke 1975).

Laboratory studies conducted at room temperature (Ewers 1930) showed that female *D. thomasi* produce 10–40 eggs per clutch. These eggs develop into adults in

TABLE 22
Reports of *Diacyclops thomasi* in the Great Lakes

	Sampling Date	Abd[a]	Reference		Sampling Date	Abd[a]	Reference
LAKE ERIE	1919–1920	F	Wickliff 1920	**LAKE MICHIGAN (continued)**	1973	A	Stewart 1974
	1928–1930	C	Wright 1955		1973–1974	A	Torke 1975
	1928	F	Sibley 1929		1974	A	Evans and Stewart 1977
	1928–1929	C	Wilson 1968		1975–1977	—	Hawkins and Evans 1979
	1929	F	Ewers 1933				
	1938–1939	A	Chandler 1940	**LAKE HURON**	1907	P	Sars 1915
	1946–1947	A	Andrews 1953		1967–1968	A	Carter 1969
	1950–1951	C	Davis 1954		1968	A	Patalas 1972
	1956–1957	A	Davis 1962		1970	A	Watson and Carpenter
	1961	A	Britt et al. 1973		1970–1971	A	Carter 1972
	1961–1962	P	Hohn 1966		1974–1975	C–A	McNaught et al. 1980
	1967	A	Davis 1968				
	1968	A	Patalas 1972	**LAKE ONTARIO**	1967	A	Patalas 1972
	1968, 1970	P	Heberger and Reynolds 1977		1967–1968	A	Patalas 1969
	1970	A	Watson and Carpenter 1974		1970	A	Watson and Carpenter 1974
	1971–1972	A	Rolan et al. 1973		1971–1972	A	Wilson and Roff 1973
	1973–1974	P	Czaika 1978b		1972	A	McNaught and Buzzard 1973
					1972	A	Czaika 1974a
LAKE MICHIGAN	1882	A	Forbes 1882		1972–1973	A	Czaika 1978b
	1887–1888	A	Eddy 1927				
	1926–1927	P	Eddy 1934	**LAKE SUPERIOR**	1889	A	Forbes 1891
	1954–1955, 1958	C	Wells 1960		1928	C	Eddy 1934
	1954–1961	F	Wells and Beeton 1963		1967–1969	C–A	Swain et al. 1970b
	1966–1967	F	Morsell and Norden 1968		1967–1968	A	Patalas 1972
	1966, 1968	C	Wells 1970		1971–1972	A	Selgeby 1975a
	1969–1970	A	Gannon 1972a		1971	A	Selgeby 1974
	1969–1970	A	Gannon 1974		1973	A	Watson and Wilson 1978
	1969–1970	A	Gannon 1975		1973	A	Upper Lakes Ref. Group 1977
	1971	A	Howmiller and Beeton 1971		1974	C	Basch et al. 1980
	1971–1972	A	Beeton and Barker 1974		1979–1980	C–A	This study
	1972–1977	C	Evans et al. 1980				

[a] Abundance Code
R = rare   U = uncommon   P = present   C = common   A = abundant   F = found in fish stomach contents
— = abundance ranking not appropriate

28–35 days. Females can produce a second clutch of eggs 4 days after the first batch hatches to the nauplius stage.

Torke (1975) also conducted laboratory experiments and found that development time shows a linear dependency on temperature. Development times for Lake Michigan field populations were identical to those of laboratory populations reared at the same temperature.

ECOLOGY IN THE GREAT LAKES

Habitat. *D. thomasi* is a cool-water species (Andrews 1953; Stewart 1974) found in lakes, ponds, and rivers throughout the year (Forbes 1897). Copepodids may be more abundant nearshore (Stewart 1974), but adults are generally found in the open water (Forbes 1897; Stewart 1974) further from shore than *Acanthocyclops vernalis* (Heberger and Reynolds 1977).

This cyclopoid is evenly distributed throughout the water column during isothermal conditions (Carter 1969; Wilson and Roff 1973) but concentrates in the upper 20m (Wells 1960) or near the thermocline (Carter 1969; Wilson and Roff 1973) during stratification. Older copepodids may occupy deeper water strata than the younger stages (Stewart 1974).

In Lake Erie *D. thomasi* moves toward the bottom during July and August when the surface waters become warmer. The low oxygen concentrations encountered in the hypolimnion may affect this species adversely and be responsible for its slight decline in Lake Erie in recent years (Wilson 1960; Heberger and Reynolds 1977).

Diurnal Migration. Adult *D. thomasi* show a regular vertical migration pattern, coming to the surface at night and returning to the lower strata during daylight hours (Wells 1960; Stewart 1974). The mean depth of this species is positively correlated with the level of incident radiation at that depth (Wilson and Roff 1973).

Feeding Ecology. *D. thomasi* has been described as a raptorial feeder (McQueen 1969). Fryer (1957a) observed that this cyclopoid was an efficient predator and often attacked prey larger than itself. It tended to nibble pieces from larger prey and several *D. thomasi* might consume the same prey.

In laboratory experiments CIV and CV copepodids selected diaptomid and cyclopoid nauplii, cyclopoid copepodids, and rotifers as prey. Few adult diaptomids or cladocerans were consumed in these studies (McQueen 1969). Later studies (Anderson 1970) found no evidence of predation on cyclopoid nauplii. Instead the experimental animals selected *Ceriodaphnia* and diaptomid stage CIV and CV copepodids. Lane (1978) found that adult *D. thomasi* readily consume diaptomids, *Daphnia galeata mendotae*, *D. longiremis*, *Bosmina longirostris*, *Tropocyclops prasinus mexicanus*, and other *Diacyclops thomasi*. The choice of prey may depend on the nutritional state of the predator and the shape, size, and swimming speed of the potential prey (Anderson 1970).

McNaught et al. (1980) studied the grazing efficiency of *D. thomasi* in Lake Huron. Filtering, ingestion, and assimilation rates were determined for animals feeding on nannoplankton and net plankton.

D. thomasi also preys on larval fish. It attacks the fish on the caudal and ventral fins and can remove pieces up to its own body size. The fish become weakened by such harassment and can become seriously injured when it cannot flick off the predators, which often attack in large numbers (Davis 1959b; Fabian 1960).

As Food for Fish. Along with diaptomids, *D. thomasi* is an important item in the diet of alewife (Morsell et al. 1968) and small bloater (Wells and Beeton 1963) in Lake Michigan. It is also consumed by ciscoes, carpsuckers, shiners, bass, perch, and young whitefish and walleye (Forbes 1882; Wickliff 1920; Sibley 1929; Ewers 1933, Wilson 1960; Hohn 1966).

LIFE HISTORY AND ECOLOGY IN OTHER LAKES

D. thomasi is most common in large, deep, clear-water lakes (Patalas 1971). It has been found to comprise up to 90% of the copepod population of some lakes (Engel 1976).

The life cycle of this species is quite variable. In Colorado it can exhibit a monocyclic, dicyclic, or acyclic pattern of abundance peaks. It is often active year-round (Armitage and Tash 1967), but under adverse conditions in some lakes it will form cysts and diapause in the CIV stage (Birge and Juday 1908; Moore 1979a; Cole 1953; Yeatman 1956; Armitage and Tash 1967; Watson and Smallman 1971; Elgmork 1973). Cysts or cocoons have been collected from Lake Mendota, Wisconsin during the fall (Birge and Juday 1908). The cysts are 0.65×0.50 mm and composed of mud and vegetable debris held together by an adhesive substance. The cyclopoid is completely concealed except for the protrusion of the setae of the caudal rami and several cocoons may be attached to each other. Cessation of diapause may occur at any time and was not attributed to a specific change in environmental conditions (Watson and Smallman 1971).

Torke (1975) was unable to find a diapause in the Lake Michigan populations of *D. thomasi* he studied. He speculated that diapause is triggered by low oxygen conditions and that such a strategy is not necessary in the oxygenated hypolimnion of Lake Michigan.

Mesocyclops edax (S. A. Forbes) 1891

TAXONOMIC HISTORY

In 1891 Forbes described *Cyclops edax* as a new species of cyclopoid copepod from North America. It was distinguished from the similar *C. leuckarti* Claus 1857 by characteristics of the hyaline plate on the terminal segment of the first antenna, the relative length of the antennal segments, and differences in setae length on the fourth and fifth legs (Forbes 1897). Marsh (1910) reexamined these characteristics and found variations in the hyaline plate of animals from the same population. He concluded that Forbes' characteristics were not sufficient to distinguish *C. edax* as a distinct species. In 1918 Sars revised the genus *Cyclops* and listed *C. leuckarti* (including the forms described by Forbes in 1891) as *Mesocyclops leuckarti*. Further work by Kiefer (1929) showed that the American specimens possess hairs on the inner margin of the rami and described them as a variety, *M. leuckarti edax*. Examination of samples from many regions of the United States allowed Coker (1943) to conclude that there were many differences between *leuc-*

karti and *edax*, and that *M. edax* was a distinct species. Yeatman (1959) and Smith and Fernando (1978) have reaffirmed these species differences.

Due to the changes in the taxonomy of this species, it has been reported from the Great Lakes as *Cyclops leuckarti*, *Cyclops leuckarti edax* and *Mesocyclops leuckarti*. Since 1960 all reports have used *M. edax*.

DESCRIPTION

S. A. Forbes (1891), E. B. Forbes (1897), Marsh (1893), and Coker (1943) detail the form of this animal. *M. edax* is distinguished from other Great Lakes cyclopoids by its four long, palmately spread, terminal setae on the caudal rami. The rami are often spread into a V and possess a row of fine hairs on their inner margins (Plate 35). Czaika (1982) provides a key to the nauplii and copepodid stages of this copepod.

SIZE

Yeatman (1959) reported that adult females ranged from 1.0–1.5 mm long and males from 0.7–0.9 mm. In Lake Superior, we found the animals to be a bit larger, with females from 1.3–1.7 mm (avg. 1.47mm) and males from 0.8–1.0 mm (avg. 0.88 mm). Comita (1972) found that nauplii grew from 0.14 mm (NI) to 0.30 mm (NV), and that the first copepodid stage was only 0.33 mm long.

DISTRIBUTION AND ABUNDANCE.

M. edax is found throughout North America, occurring as far south as Mexico. It is much more common than *M. leuckarti* in the United States (Coker 1943).

While occurring in all of the Great Lakes (Table 23), *M. edax* is most common in Lake Erie and Lake Michigan's Green Bay where peak abundance can reach 22000 and

TABLE 23
Reports of *Mesocyclops edax* in the Great Lakes

	Sampling Date	Abd[a]	Reference		Sampling Date	Abd[a]	Reference
LAKE ERIE	1928–1929	C	Wilson 1960	**LAKE MICHIGAN (continued)**	1971–1972	U	Beeton and Barker 1974
	1938–1939	C	Chandler 1940		1972–1977	P	Evans et al. 1980
	1946–1948	C	Andrews 1953		1973	R	Stewart 1974
	1950–1951	C	Davis 1954				
	1956–1957	C	Davis 1962	**LAKE HURON**	1907	C	Sars 1915
	1958	P	Davis 1959a		1967–1968	U	Patalas 1972
	1961	P–C	Britt et al. 1973		1967–1968	C	Carter 1969
	1967	P–C	Davis 1968		1970	U	Watson and Carpenter 1974
	1967–1968	P	Davis 1969				
	1967–1968	P	Patalas 1972		1970–1971	C	Carter 1972
	1968, 1970	P	Heberger and Reynolds 1977		1974–1975	P	McNaught et al. 1980
	1970	C	Watson and Carpenter 1974	**LAKE ONTARIO**	1967	P	Patalas 1969
	1971–1972	C	Rolan et al. 1973		1967–1968	U	Patalas 1972
	1973–1974	P	Czaika 1978b		1970	U	Watson and Carpenter 1974
LAKE MICHIGAN	1954–1961	PF	Wells and Beeton 1963		1972	R	Czaika 1974a
	1954–1955, 1958	C	Wells 1960		1972	P	McNaught and Buzzard 1973
	1966, 1968	R–U	Wells 1970				
	1969–1970	P	Gannon 1972a	**LAKE SUPERIOR**	1967–1968	U	Patalas 1972
	1969–1970	P	Gannon 1974		1971–1972	P	Selgeby 1975a
	1969–1970	U	Gannon 1975		1971	C	Selgeby 1974
	1971	P	Howmiller and Beeton 1971		1973	U	Upper Lakes Ref. Group 1977
					1979–1980	P	This study

[a] Abundance Code
R = rare   U = uncommon   P = present   C = common   A = abundant   F = found in fish stomach contents

99

1400 animals m⁻³ respectively. Wells (1970) reported that this species was more important in Lake Michigan before the introduction of alewife.

LIFE HISTORY IN THE GREAT LAKES

M. edax is generally absent from samples collected from late fall until early spring (Andrews 1953; Davis 1962). During the summer, this species increases in abundance, and population peaks occur between July and September (Chandler 1940; Davis 1954; Wells 1960; Wilson 1960; Carter 1972; Britt et al. 1973; Watson and Carpenter 1974; Heberger and Reynolds 1977).

The life history of this cyclopoid copepod is best described by Selgeby (1975) for Lake Superior populations. He first collected the species in June and July, when all specimens were stage CV copepodids. They matured by the end of July, with CVI females outnumbering males 12:1. Reproductive activity was high at this time as indicated by the presence of spermatophores or egg sacs on 100% of the females. Clutch size averaged 19 eggs 109μm in diameter. The eggs developed fairly rapidly, with copepodid stages CI and CII occurring by mid August. By late August 50% of this generation had matured, with the female to male ratio now 3:1. Females mated and produced an average of 16 eggs with a mean diameter of 105μm. By October this second generation had reached copepodid stages CIII and CIV. They later molted to stage CV and diapaused near or in the bottom sediments until spring.

The adult sex ratio is quite variable. Andrews (1953) found a female to male ratio of 2.2:1, Davis (1968) found slightly more females than males, and Davis (1962) found more males than females during the summer population peak.

ECOLOGY IN THE GREAT LAKES

Habitat.   The distribution of this species is related to water temperature. In Lake Erie, which is fairly warm, greatest abundances are found throughout the littoral zone and near the bottom (Wilson 1960). The other lakes are cooler; the animals often concentrate in the warm epilimnion of Lake Michigan (Wells 1960) and in the nearshore regions of Lake Huron (Carter 1972).

Andrews (1953) found that *M. edax* breaks diapause when water temperature reaches 8°C. The population peaks when temperature reaches 22°C and begins to disappear when it cools to 13°C.

Diurnal Migration.   Although Wells (1960) noted an increased abundance of this species at the surface at night, the extent of the migration is not known.

Feeding Ecology.   *M. edax* is predaceous, known to attack fish fry near the posterior fins and tail, often seriously injuring them (Davis 1959b).

Several laboratory and field studies have studied the feeding behavior of this species. Prey choice and the capture success of *M. edax* depends on the probability of encountering an item, the attack threshold of *M. edax*, the probability of a successful seizure, and the types of prey present (Brandl and Fernando 1975). Small prey, including rotifers and small *Daphnia*, are preferred (Brandl and Fernando 1978, 1979). The type of movement and body structure of the rotifers were found to be more important than body size in determining which ones would be eaten.

M. edax has also been found to consume *Diaphanosoma*, *Daphnia*, cyclopoid and calanoid copepodids, and copepod nauplii (Fryer 1957b; Smyly 1961a; and Lane 1978). In laboratory experiments outside the Great Lakes area, Confer (1971) found that *M. edax* selectively preyed upon *Diaptomus floridanus* even though cladocerans were present in greater numbers. The predation pressure on *D. floridanus* (an herbivore) was considered sufficient to greatly influence the species and size composition of the entire herbivore population.

The lack of green coloration in the gut of *M. edax* collected from field samples led Confer (1971) to believe that this species was a strict carnivore during his study period. Karabin (1978) found that in lakes in Poland and Russia, *M. edax* consumed rotifers, nauplii, protozoans, bacteria, detritus, and algae. In Lake Huron *M. edax* filters nannoplankton at a rate of 0.058−0.088 ml·animal⁻¹·hour⁻¹ and consumes 12.8% of its body weight per day (McNaught et al. 1980).

As Food for Fish.   *M. edax* is consumed by white bass, largemouth bass, bullhead, small bloater, crappie, freshwater drum, yellow perch, and other small fish (Wickliff 1920; Sibley 1929; Ewers 1933; Wells and Beeton 1963; Wilson 1960). Wells (1960) observed that the abundance of this species in Lake Michigan declined sharply from 1954 to 1966, the same time that the planktivorous alewife population increased markedly.

LIFE HISTORY AND ECOLOGY IN OTHER LAKES

This species is common to small lakes and ponds. In most areas it diapauses during the winter as copepodid stage CIV or CV. Diapause may be initiated by decreasing photoperiod (Armitage and Tash 1967) and usually lasts 5 months. Ovigerous females appear in April or May and produce two generations each year, similar to the reproductive pattern of the Great Lakes (Smyly 1961b; Armitage and Tash 1967; Comita 1972).

The pattern of diapause in meromictic lakes is similar to that of most other lakes, but since the deep layers of meromictic lakes do not mix and have constant light, temperature, and oxygen conditions, some investigators suspect that diapause in these waters is controlled by an innate physiological process independent of the environment (Smyly 1961a, 1961b; Elgmork 1973).

In lake populations it takes *M. edax* 70 days to develop from eggs laid in June to the adult stage; 48 days for naupliar development, and 22 for copepodid development (Smyly 1961a). Development rate is temperature dependent (Gophen 1976). In laboratory experiments at 14–18°C, it only took 16 to 25 days for development from egg to adult. The adults lived 40–45 days, producing 5–8 clutches of eggs at intervals of 4–5 days (Ewers 1930, Smyly 1961b). Detailed descriptions of egg hatching are given by Davis (1959a).

Studies on the seasonal abundance of populations of *M. edax* and *Bosmina longirostris* (DeCosta and Janicki 1978) and *M. edax* and *Diacyclops thomasi* (Armitage and Tash 1967) have been reported.

Tropocyclops prasinus mexicanus Kiefer 1938

TAXONOMIC HISTORY

Like most other cyclopoids, the taxonomic status of *Tropocyclops prasinus* (Fischer) 1860 has changed considerably since it was first described. For many years this species was considered a member of the *Cyclops serrulatus* group and was placed in the subgenus *Eucyclops* (Kiefer 1927). Gurney (1933) held that it was morphologically distinct from the other *Eucyclops* species and was similar in many ways to *Macrocyclops*. He continued to call it *Cyclops prasinus* but placed it in a new subgenus, *Tropocyclops*. When the taxonomy of the cyclopoids was revised, *Tropocyclops* was elevated to genus level (Kiefer 1960; Dussart 1969). In 1938, Kiefer described another form of *T. prasinus* that he named *T. prasinum mexicanus*. Only the subspecies *mexicanus* has been identified from the Great Lakes region (Robertson and Gannon 1981). Therefore, all reports of *Eucyclops prasinus* and *Tropocyclops prasinus* from the Great Lakes are included in this discussion of *T. prasinus mexicanus*.

DESCRIPTION

Tropocyclops prasinus mexicanus is a very small cyclopoid with 12-segmented first antennae reaching to the end of the metasome. The urosome is narrow and elongate. Adult males are distinguished from females by their geniculate first antennae (Plate 36). More detailed descriptions of this species are given by Gurney (1933) and Yeatman (1959). Czaika (in press) includes a key to the nauplii and copepodid stages.

SIZE

Adult females range from 0.50–0.90 mm long while males only reach 0.5–0.6 mm (Forbes 1897; Gurney 1933; Yeatman 1959). In Lake Superior, we found females with an average length of only 0.57 mm. In Lake Michigan copepodid stages CI–CV have dry weights of 0.3–0.5μg while CVI adults weigh 0.7–1.2μg (Hawkins and Evans 1979).

DISTRIBUTION AND ABUNDANCE

T. prasinus has a worldwide distribution, occurring in Europe, Asia, Africa, Australia, and North and South America. In the United States *T. prasinus mexicanus* is common in the Great Lakes region.

This subspecies has been reported from Lakes Erie and Michigan since the early 1900s and from Lakes Huron and Ontario since 1967 (Table 24). It was found in a few samples from Lake Superior in 1974 (Basch et al. 1980), and we collected it in several samples from the Duluth-Superior area in 1979 and 1980. *Tropocyclops* species are usually found at densities less than 100 m^{-3} (McNaught and Buzzard 1973; Stewart 1974), but peaks of 200–4000 m^{-3} have been reported (Howmiller and Beeton 1971; Rolan et al. 1973; Watson and Carpenter 1974). Some of the variability in abundance reports may be due to differences in mesh size of the nets used. Carter (1969) found that this small cyclopoid can escape nets with a mesh greater than 160μm

LIFE HISTORY IN THE GREAT LAKES

In the Great Lakes, *T. prasinus mexicanus* is found throughout the year (Davis 1962; Beeton and Barker 1974; Stewart 1974; Watson and Carpenter 1974; Gannon 1975; Czaika 1974a, 1978b). Abundance peaks between August and November (Chandler 1940; Wells 1960; Patalas 1969; Carter 1969, 1972; Wilson and Roff 1973; Torke 1975; Czaika 1978b), and then declines through the winter (Rolan et al. 1973; Stewart 1974).

Torke (1975) studied the reproduction of *T. prasinus mexicanus* in Lake Michigan. Breeding was restricted to the period of thermal stratification from June until October. Reproduction was rapid and continuous, with the population size increasing logarithmically. Reproduction declined in the fall, and the animals overwintered in the adult stage.

TABLE 24
Reports of *Tropocyclops prasinus mexicanus* in the Great Lakes

	Sampling Date	Abd[a]	Reference		Sampling Date	Abd[a]	Reference
LAKE ERIE	1928–1930	P	Wright 1955	**LAKE MICHIGAN (continued)**	1973–1974	P	Torke 1975
	1928–1929	P	Wells 1960		1974	P	Evans and Stewart 1977
	1938–1939	P–C	Chandler 1970		1975–1977	—	Hawkins and Evans 1979
	1956–1957	U	Davis 1962				
	1961	U	Britt et al. 1973	**LAKE HURON**	1967–1968	U	Patalas 1972
	1967–1968	U	Patalas 1972		1967–1968	P–C	Carter 1969
	1967–1968	U	Davis 1969		1970	U–P	Watson and Carpenter 1974
	1968, 1970	U	Heberger and Reynolds 1977		1970–1971	P–C	Carter 1972
	1970	U	Watson and Carpenter 1974		1974	P	Basch et al. 1980
	1971–1972	P	Rolan et al. 1973		1974–1975	P–C	McNaught et al. 1980
	1973–1974	P	Czaika 1978b				
				LAKE ONTARIO	1967	C	Patalas 1969
LAKE MICHIGAN	1887–1888	C	Eddy 1927		1967–1968	U	Patalas 1972
	1926–1927	A	Eddy 1927		1969, 1972	P	McNaught and Buzzard 1973
	1954–1955, 1958	P	Wells 1960		1970	P–C	Watson and Carpenter 1974
	1969–1970	P	Gannon 1972a		1971–1972	C	Wilson and Roff 1973
	1969–1970	P	Gannon 1974		1972	C	Czaika 1974a
	1969–1970	P	Gannon 1975		1972–1973	P–C	Czaika 1978a
	1971	P	Howmiller and Beeton 1971				
	1971	P–C	Beeton and Barker 1974	**LAKE SUPERIOR**	1973	U	Upper Lakes Ref. Group 1977
	1972–1977	P	Evans et al. 1980		1974	R	Basch et al. 1980
	1973	P	Stewart 1974		1979–1980	U	This study

[a] Abundance Code
R = rare   U = uncommon   P = present   C = common   A = abundant   F = found in fish stomach contents
— = abundance ranking not appropriate

This cycle is supported by several other studies (Ewers 1936; Carter 1969; Gannon 1972a; Stewart 1974), but the number of clutches per female and number of generations produced each year have not yet been determined.

Adult females usually outnumber males (Torke 1975) except on one occasion in October when Stewart (1974) found males to be more numerous.

ECOLOGY IN THE GREAT LAKES

Habitat.   *T. prasinus mexicanus* is usually found in small lakes. In the Great Lakes it prefers warm water masses (Patalas 1969) and concentrates in the surface waters above 3 m during the summer and fall (Carter 1969; Wilson and Roff 1973). The animals are more evenly dispersed in the water column during the winter (Wilson and Roff 1973). This cyclopoid usually has a uniformly sparse horizontal

distribution (Howmiller and Beeton 1971) but occasionally occurs in large patches (Gannon 1975).

Diurnal Migration.   This cyclopoid does not exhibit a regular migration pattern. Its vertical distribution varies and does not correlate with surface temperature, chlorophyll concentration, or radiation intensity (Wilson and Roff 1973).

Food and Feeding Behavior.   *T. prasinus mexicanus* is omnivorous. Predation experiments conducted by Lane (1978) showed that it consistently preyed upon *Daphnia longiremis* and diaptomid copepodids. It consumed *Daphnia galeata*, *D. retrocurva*, *D. dubia*, *Diaphonosmoa leuchtenbergianum*, *Bosmina longirostris* and *Diacyclops thomasi* less often. This subspecies also exhibits cannibalism. Lane concluded that it was a more effective predator on small cladocerans than *Diacyclops thomasi*. McNaught et al. (1980) found that it filters nanoplankton in Lake Huron at a rate of 0.002–0.043 ml·animal⁻¹·hour⁻¹ and ingests an average of 1.4% of its body weight each day.

As Food for Fish and Other Organisms. There are no reports of fish consuming *T. prasinus mexicanus* in the Great Lakes. However, Lane (1978) showed *T. prasinus mexicanus* is readily captured and consumed by *Diacyclops thomasi*.

LIFE HISTORY AND ECOLOGY IN OTHER LAKES

In Lake Cromwell, Quebec *Tropocyclops prasinus* is multivoltine and produces two generations each year. The first generation matures in 67 days between July and September. The second generation hatches between mid-August and October and grows to the CV and CVI copepodid stage before overwintering. This second generation requires approximately 195 days to mature and doesn't reproduce until June or July (Paquette and Pinel-Alloul 1982).

Ewers (1930) studied the development of *T. prasinus mexicanus* in the laboratory. Females produced 12–14 ova in each of their two egg sacs, and 2–3 days after the first brood hatched, produced two more egg sacs. At room temperature, development from egg to adult was rapid, requiring only 17 days.

Tropocyclops prasinus is usually found in lakes with a pH of 6.1–7.6 (Ward 1940) but can withstand more basic pH conditions (7.0–8.46) in the lower Atchafalya River Basin, Lousiana (Binford 1978).

MALACOSTRACANS

Mysis relicta (Lovén) 1861

TAXONOMIC HISTORY

It is generally considered that *Mysis relicta* developed from brackish water populations of *M. oculata* isolated in fresh water after the Pleistocene glaciers. *M. relicta* has occasionally been classified as a subspecies or variety of *M. oculata* (Juday and Birge 1927; Creaser 1929) but is readily distinguished from the latter and is now recognized as a distinct species (Tattersall and Tattersall 1951). This species is the only representative of the Mysidacea found in the Great Lakes (Holmquist 1972) and was probably misidentified in 1872 when Hoy reported finding *Mysis delurranius* in Lake Michigan.

DESCRIPTION

The mysids, or opossum shrimp, are large transparent zooplankton with stalked eyes. Mature females are distinguished by plates on the last two thoracic segments, which are used to enclose developing young in the brood pouch (Plate 37) (Larkin 1948). Mature males lack these plates but have greatly enlarged fourth pleopods (Plate 38). The sexes can be distinguished by the shape of the third and fourth pleopods, which are biramous in males, when the young reach a length of 6 mm. A more detailed description of *Mysis relicta* is given by Tattersall and Tattersall (1951).

SIZE

Juvenile mysids in Lake Michigan range from 3–13 mm long. Adult males are generally smaller (13–21 mm) than adult females (14–25 mm) (Reynolds and DeGraeve 1972).

DISTRIBUTION AND ABUNDANCE

M. relicta, a relict species with a holarctic distribution, is found throughout the formerly glaciated areas of North America and Europe (Holmquist 1972; Segersträle 1962).

The species also occurs in brackish waters along the coast of Greenland but is absent from the more saline open waters of the Arctic Ocean (Holmquist 1963; Johnson 1964; Dadswell 1974). Recently this species has been introduced into cold, deep-water lakes in Sweden (Furst 1981) and in Lakes Kalamalda and Pinaus in British Columbia (Stringer 1967), Grindstone Lake in Minnesota (Schumacher 1966), and Lake Tahoe, California-Nevada (Linn and Frantz 1965).

M. relicta has been observed in all five Great Lakes (Table 25) but is best known from Lake Michigan. In Lake Erie this species is generally restricted to the deeper, colder waters of the eastern basin (Wilson 1960; Carpenter et al. 1974). In the deeper waters of Lake Huron's central and northern basins, this mysid has an average abundance of 60 m⁻² and exhibits peaks of 847 m⁻². Lakes Michigan, Superior, and Ontario have larger populations (averaging 112–188 m⁻²) with peaks of 800–1000 m⁻² (Grossnickle and Morgan 1979; Carpenter et al. 1974; Morgan and Beeton 1978)

LIFE HISTORY IN THE GREAT LAKES

The life history of *M. relicta* is best known from studies conducted on Lake Michigan (McWilliam 1970; Reynolds and DeGraeve 1972; Mozley and Howmiller 1977; Morgan and Beeton 1978). In the shallower areas, mysids live approximately one year. During the winter, adults may move inshore (27–55 m) and mating is observed. Males die soon

TABLE 25
Reports of *Mysis relicta* in the Great Lakes

	Sampling Date	Abd[a]	Reference		Sampling Date	Abd[a]	Reference
LAKE ERIE	1919–1920	F	Clemens and Bigelow 1922	**LAKE MICHIGAN (continued)**	1973–1974	P	Torke 1975
	1928–1929	P–C	Wilson 1960		1975–1976	C	Morgan and Beeton 1978
	1971	P	Carpenter et al. 1974		Uncertain	P	Mozley and Howmiller 1977
	1979	R	Fitzsimons 1981		1976	C	Grossnickle and Morgan 1979
LAKE MICHIGAN	1870	P	Stimpson 1871				
	1870	P	Hoy 1872	**LAKE HURON**	1955–1956	C	Beeton 1960
	1872	F	Smith 1874a		1971	C–A	Carpenter et al. 1974
	1925	F	Creaser 1929		1974	P	Basch et al. 1980
	1931–1932	P	Eggleton 1936				
	1932	F	Van Oosten and Deason 1938	**LAKE ONTARIO**	1919–1920	F	Clemens and Bigelow 1922
	1933–1936	F	Schneberger 1936		1927–1928	F	Pritchard 1929
	1954–1955	F	Bersamin 1958		1967	P	Patalas 1969
	1954–1955	P	Beeton 1960		1971–1972	C	Wilson and Roff 1973
	1954–1955, 1958	P	Wells 1960		1971	C–A	Carpenter et al. 1974
	1954–1961	F	Wells and Beeton 1963				
	Uncertain	A	Anderson 1967	**LAKE SUPERIOR**	1870	P	Smith 1871
	1965	P	McNaught and Hasler 1966		1872	F	Smith 1874a
	1966–1967	F	Morsell and Norden 1968		1933–1934	P	Eddy 1943
	1967	A	Robertson et al. 1968		1950–1953	F	Dryer et al. 1965
	1970–1971	C	Reynolds and Degraeve 1972		1966–1968	F	Bailey 1972
	1971–1972	P	Beeton and Barker 1974		1967–1969	P	Swain et al. 1970b
	1972–1977	P	Evans et al. 1980		1971	C–A	Carpenter et al. 1974
					1974	P	Basch et al. 1980
					1979–1980	P	This study

[a] Abundance Code
R = rare   U = uncommon   P = present   C = common   A = abundant   F = found in fish stomach contents

after mating, but the female broods the eggs for approximately 3–4 months. When the water begins to warm in April or May, she releases the fully developed young and moves offshore to cooler waters.

In deeper waters of Lake Michigan (> 54 m) mysids breed more continuously, with peaks of activity in April and May. Upon reaching a length of 16 mm, the female deposits eggs into the brood pouch and mates. At this size, the female carries 17 or 18 eggs, but larger broods are common in larger animals. Females continue to grow 1 mm per month while the embryos develop and may be 22 mm long when the young are released after 5 months of incubation.

Some mortality occurs in the brood pouch as indicated by a decrease in numbers as the eggs develop into emybryos and larvae. It is not yet known whether the dead are aborted or reabsorbed by the females.

Lake Michigan mysids generally have four developmental instars. Stage I is 4.5 mm long, stage II 9 mm, stage III 12 mm, and the sexually mature stage IV measures 16 mm. During July and August approximately 6% of the females will molt to a fifth instar stage and produce a second brood before dying, while males die after stage IV. (Morgan and Beeton 1978). The female to male ratio is 1:1 when immature stages are considered, but jumps to 5:1 for adults during the breeding season (Morgan and Beeton 1978).

The length of the life cycle and season of peak reproductive activity differ in the other Great Lakes (Carpenter et al. 1974). In Lakes Ontario and Huron mysids may live 18 months, and young are recruited into the population throughout the summer and into October. In Lake Superior mysids live 2 years, and major recruitment occurs from early spring until July. Two major size classes are present year-round in Lake Superior. Mysids in the deepest portions of Lake Michigan are also thought to have a 2-year life span (Mozley and Howmiller 1977). These findings agree with Lasenby and Langford's findings (1972, 1973) that *M. re-*

licta has a higher metabolic rate in warm, productive lakes and can mature in 1 year while 2 years are required for maturation in colder, arctic lakes.

ECOLOGY IN THE GREAT LAKES

Habitat. *Mysis relicta* is found year-round in deep, cold well-oxygenated areas of the Great Lakes. It prefers temperatures of 1–6°C and is rarely found in water depths less than 26 m during the summer (Reynolds and DeGraeve 1972; Carpenter et al. 1974). However, during the winter and spring, before the water mass warms, mysids may be found inshore in water only 6–18 m deep (Reynolds and DeGraeve 1972; Mozley and Howmiller 1977).

Mysids are more abundant in the deeper areas of all the Great Lakes and may concentrate near the base of steep slopes (Anderson 1967). In very deep stations (260 m) in Lake Michigan, Robertson et al. (1968) found two separate concentrations of mysids. One group was active at a depth of 187 m and the other remained near the bottom. Due to Lake Erie's warm temperatures and low hypolimnetic oxygen levels, mysids are generally restricted to the deep holes in the eastern basin during the summer (Beeton 1965).

Diurnal Migration. *M. relicta* is well known for its extensive vertical migrations. It is very sensitive to light (especially to wavelengths near 515 nm) and, in waters less than 100-m deep, rests on or near the bottom during the day. As light intensity drops at sunset, migration toward the surface begins at rate of 0.1–0.8 m/min (6–48 m/hr). This ascent is slowed or reversed if the animals encounter higher light intensities near the surface due to moonlight (Beeton 1960; McNaught and Hasler 1966). About an hour before sunrise, mysids begin their descent to daytime habitats at rates of 5–24 m/hr.

Immature mysids may rest above the adults during the day and have been found to ascend higher in the water column at night (Beeton 1960; McWilliams 1970).

Low water temperatures do not affect *M. relicta*'s migration patterns or speed but warm-water masses form a barrier to migration. In August, 75% of Lake Michigan mysids did not cross into the thermocline, instead forming a band near its lower limit (McNaught and Hasler 1966).

Wilson and Roff (1973) found that seasonal changes in the migration pattern of mysids in Lake Ontario was correlated with the maximum daytime light intensity and the surface concentration of chlorophyll.

All mysids do not exhibit the same pattern of resting near the bottom during the day and moving toward the surface at night. At depths over 150 m, where light intensities remain low all day, some mysids remain pelagic all day while others tend to remain near the bottom and may not migrate at all (Robertson et al. 1968).

Food and Feeding Behavior. Mysids are generally filter feeders, straining suspended particles out of the water with their inner thoracic appendages. In laboratory studies they were observed swimming on their backs to obtain food particles from the water surface (DeGraeve and Reynolds 1975). Mysids also stir up bottom sediments to obtain settled particles and may seize other zooplankters.

Analyses of stomach contents by Forbes (1882) and Smith (1874a) revealed that *Bosmina, Daphnia, Canthocamptus* and cyclopids were consumed by mysids in the Great Lakes. More recently (McWilliam 1970) mysids were found to be "facultative herbivores," feeding on algae, detritus, cladocerans, copepods and juvenile mysids. Food selection was based on the relation of particle size to body size.

In laboratory feeding studies (Grossnickle 1978; Bowers and Grossnickle 1978), *M. relicta* instars I and II from Lake Michigan consumed the 10–53μm and the > 53μm size fractions of the natural phytoplankton mixture they were offered. Adults generally consumed the > 53μm size class and exhibited higher filtering rates when fed at warmer temperatures approximating those of the thermocline.

Field sampling (Grossnickle 1978) confirmed that Lake Michigan mysids are primarily herbivores during their summer vertical migrations, and feeding on algae increased when they came to the surface at sunset. However, mysids also fed readily on zooplankton when these items were offered in the lab. Prey consumption depended on the prey's ability to escape (Grossnickle 1978).

Recent reports by Parker (1980), Bowers and Vanderploeg (1982), Grossnickel (1982), Vanderploeg et al. (1982), Mederia et al. (1982), Sierszen and Brooks (1982), Beeton and Bowers (1982), Sell (1982), Seale and Boraas (1982), and Evans et al. (1982) discuss the predatory behavior, nutrient excetion, vertical migration and PCB concentration in *M. relicta* from Lake Michigan.

As Food for Fish. Mortality of *M. relicta* is very high in Lake Superior as indicated by the large percentage of juveniles in the population (Carpenter et al. 1974). Mysids are a major food item of essentially all fish occupying deeper areas of the Great Lakes, including smelt and alewife greater than 10 cm, ciscoes, bloater, burbot, young lake trout, and fourhorned sculpin (Morsell and Norden 1968; Anderson and Smith 1971; Reynolds and DeGraeve 1972). Ninety percent of whitefish examined by Smith (1974a) and 59% of the larger bloaters studied by Wells and Beeton (1963) contained mysids.

ECOLOGY AND LIFE HISTORY IN OTHER LAKES

Berrill (1969) conducted laboratory studies on the embryonic development of *M. relicta* from Stony Lake in Ontario, Canada. He found that the behavioral patterns of the embryos were integrated and coordinated at an early stage.

Reproduction in Green Lake, Wisconsin was studied by Juday and Birge (1927). Female mysids matured at 1 year, bred from October to May, and brooded an average of 19–20 eggs, although one animal was found to carry 34 young. Some females produced a second brood and lived for two years, similar to mysids in the cooler areas of the Great Lakes. Green Lake mysids were slightly larger than those of the Great Lakes. Males were 18–24 mm long and females ranged from 22–30 mm.

The British mysids (Tattersall and Tattersall 1951) are also winter breeders with egg laying commencing when water temperatures cool to 7°C. The eggs are generally 0.56–0.76 mm in diameter and larger females (17–21 mm long) may carry twice as many eggs (25–40) as the smaller ones (13–15 mm long with 10–20 eggs).

M. relicta is a cold-water species, generally restricted to lakes that do not warm above 14–16°C (Tattersall and Tattersall 1951; Dadswell 1974). Laboratory studies found that mortality increases at temperatures greater than 10–13°C (Smith 1970; DeGreave and Reynolds 1975) and an upper lethal temperature of 20.3°C was determined for organisms acclimated to 2–17°C (Dadswell 1974).

Smith (1970) also found that mysids are very light-sensitive and cannot tolerate normal laboratory illumination. Although not affected by pH in the range of 6–8 (Dadswell 1974), mysids are restricted to areas with hypolimnetic oxygen concentrations of at least 1–4 cc/l (Tattersall and Tattersall 1951).

In eutrophic Stony Lake, Ontario, Lasenby and Langford (1973) found that mysids were benthic detritivores during the day but fed on *Daphnia* at night. Those in oligotrophic Char Lake, N.W.T. were basically herbivores, although chironomids were consumed in laboratory experiments.

Other studies by Rieman and Falter (1981), Kinsten and Olsen (1981), and Lasenby and Furst (1981) document the effect of *M. relicta* predation on other zooplankton populations.

Pontoporeia Krøyer 1842

TAXONOMIC HISTORY

The first specimens of *Pontoporeia* collected from the Great Lakes (Smith 1874a) were described as two species, *P. hoyi* and *P. filicornis*. Additional specimens of *Pontoporeia* collected by Juday and Birge (1927) were judged similar to the European *P. affinis* Lindstrom 1855 and were synonymized with this species. *P. filicornis*, as redescribed by Adamstone (1928) from Lake Nipigon, was later considered synonymous with the adult male of *P. affinis* (Segersträle 1937).

Although more recent studies refer to the Great Lakes *Pontoporeia* as *P. affinis*, Segersträle (1971a) distinguished three forms of adult males in North American populations, primarily differing in their antennal length. The forms were designated *P. affinis* form *filicornis*, *P. affinis*, f. *brevicornis*, and *P. affinis* f. *intermedia*.

E. L. Bousfield and G. S. Thurston are currently revising the holarctic genus *Pontoporeia* (E. L. Bousfield pers. comm. 1983). Detailed morphological evidence reveals several distinct members in North America, all distinct from the European fresh-water members at the generic or subgeneric level. For the present, Bousfield (pers. comm. 1983) suggests that the North American species may be arranged conveniently into a "long-horned" *filicornis* group (deeper, colder waters) and a "short-horned" *brevicornis* group occurring more frequently in shallower, warmer water. These distinctions are based on the pelagic adult male morphology. In the "short-horned" form, the flagellae of both pairs of antennae have about 11–20 segments and both pairs of antennae are much shorter than the body. In the "long-horned" form, the flagellae have up to 50 or 60 segments and both pairs of antennae are longer than the body. In our discussion we will refer to the genus level only.

DESCRIPTION

Except for the adult male, *Pontoporeia* are burrowing, fossorial amphipods (Plates 39, 40). The pelagic male form is distinguished from other Great Lakes amphipods (such as *Gammarus*, *Crangonyx*, and *Hyallela*) by the elongate filiform antennae bearing calceoli (vibration sensory organs), the fossorial form of the appendages (segments broad, very setose, and spinose margins), and the presence of paired fingerlike "gills" on the sternum of some thoracic segments.

More complete descriptions of Great Lakes *Pontoporeia* are given in Segersträle (1937, 1971a), and Bousfield (1982) gives a description of the family Pontoporeidae.

SIZE

Juday and Birge (1927) found adult females 7.0–8.5 mm long in Green Lake, Wisconsin. In Great Slave Lake (Moore 1979b), adult females ranged from 6.0–9.0 mm, and March young measured 1.5 mm.

TABLE 26
Reports of *Pontoporeia* in the Great Lakes

	Sampling Date	Abd[a]	Reference		Sampling Date	Abd[a]	Reference
LAKE ERIE	1928–1929	C	Wilson 1960	**LAKE HURON**	1953, 1956	A	Teter 1960
	1974	R	Barton and Hynes 1976		1955, 1956	A	Henson 1970
					1974	C	Barton and Hynes 1976
LAKE MICHIGAN	1931–1932	A	Eggleton 1936				
	1932	F	Van Oosten and Deason 1938	**LAKE ONTARIO**	1964	P	Johnson and Mathison 1968
	1954–1955, 1958	P	Wells 1960		1966–1968	C	Johnson and Brinkhurst 1971
	1954–1955	F	Bersamin 1958		1974	R	Barton and Hynes 1976
	1954–1961	F	Wells and Beeton 1963				
	1961	C–A	Marzolf 1965a	**LAKE SUPERIOR**	1933–1934	C	Eddy 1934
	1963	P	Marzolf 1965b		1950, 1953, 1963	F	Dryer et al. 1965
	1964	A	Robertson and Alley 1966		1959–1961	A	Thomas 1966
	1964	A	Wells 1968		1966–1968	F	Bailey 1972
	1965	C	McNaught and Hasler 1966		1974	C	Barton and Hynes 1976
	1967	C	Powers and Alley 1967				
	1966–1967	F	Morsell and Norden 1968				
	1973–1974	C	Torke 1975				

[a] Abundance Code
R = rare   U = uncommon   P = present   C = common   A = abundant   F = found in fish stomach contents

DISTRIBUTION AND ABUNDANCE

Members of the genus *Pontoporeia* (as presently understood) occur in coastal marine, brackish, and fresh waters of the holarctic region. The *P. femorata* group is strictly marine, the *P. affinis* group occurs in brackish and fresh waters mainly in Europe, and the *P. hoyi* group is endemic to North American glacial relict lakes (Bousfield pers. comm. 1983).

Pontoporeia species have been found in all the Great Lakes (Table 26) although members are restricted mainly to deeper portions of the eastern basin of Lake Erie (Wilson 1960). They are abundant in Lake Superior, the midlake region of Lake Huron, and Lake Michigan, where dredge samples have collected 5000 organisms per square meter (Teter 1960; Mozley and Howmiller 1977). The genus is widely distributed in smaller glacial relict lakes of northeastern North America (Dadswell 1974).

LIFE HISTORY IN THE GREAT LAKES

In shallow water, males and females molt to the adult stage and mate in late autumn or early spring (Teter 1960; Mozley and Howmiller 1977). Mating is believed to be synchronized by a decreasing photoperiod (Segersträle 1971c) and with the fall overturn (Dadswell 1974). In water deeper than 16 m, the light cue is reduced and breeding may take place later in the year or irregularly throughout the year including the summer months (Segersträle 1967, 1971b; Alley 1968; Mozley and Howmiller 1977). Males die soon after mating, but females return to the bottom, brood the eggs for several months, then die after releasing the 2-mm young.

In shallow water, *Pontoporeia* matures in 1 year, but populations in deeper water require 2 or more years to mature (Alley 1968; Mozley 1974). These populations exhibit two distinct size classes representing organisms that mature and breed in alternate years (Alley 1968). This pattern may reflect the life histories of 2 or more species.

Nearshore population numbers increase greatly in the spring when young are released. However, mortality and emigration reduce numbers to the winter level by late fall. In deeper waters little or no seasonal change in abundance is noted due to variable breeding or migration patterns (Alley 1968; Mozley 1974).

Production and abundance of *Pontoporeia* varies widely in consecutive years; abundance measurements at a given location and time may not be representative of the entire population and therefore should be used cautiously in assessing environmental quality (Mozely and Howmiller 1977).

ECOLOGY IN THE GREAT LAKES

Habitat. *Pontoporeia* are benthic infaunal organisms that generally burrow in fine bottom sediments. Only during its brief pelagic mating forays is it taken in plankton samples, and then mainly at night or from near the bottom. Because of deficiencies in species separation, much of the ecological information in the literature is difficult to interpret (Bousfield, pers. comm. 1983). Abundance can not be well correlated with water depth (Teter 1960; Johnson and Matheson 1968; Henson 1970; Barton and Hynes 1976) but seems to depend on substrate type. The various species occur mainly in sediments with a mean grain size of less than 0.5 mm and consisting of 60–70% sand, 10–20% silt, and 10–30% clay (Marzolf 1965b; Henson 1970). They tend to avoid areas of hard rock, hard Valders red clay deposits, and regions of wave action and strong bottom currents. *Pontoporeia* are rare or lacking in deep basins of the open lake where food (detritus) is scarce or in highly acidic waters (Dadswell 1974).

Pontoporeia was formerly believed to be temperature limited and did not reproduce in the lab at temperatures greater than 7°C (Smith 1972). Thomas (1966) found fewer animals in the warmer bays of Lake Superior than in cooler areas, and *Pontoporeia* was only found at the surface of Lake Michigan in June and November when surface water temperatures were less than 11°C (Wells 1960). Recent laboratory studies on Lake Superior animals (Smith 1972) established an LD50 (24 hr) of 12°C and an LD50 (96 hr) of 10.8°C for animals acclimated to 6°C. Organisms acclimated to 9°C exhibited an LD50 (30 day) of 10.4°C, which was thought to represent the upper lethal temperature on a continuous basis. However, *Pontoporeia* in Lake Michigan have been found to persist in moderate densities in areas with constant temperatures near 19°C and in other areas with rapidly fluctuating temperatures reaching 23°C (Alley 1968, Mozley 1974). These findings suggest that temperature may be secondary to substrate in determining distribution, although all this work is suspect because the species were not accurately determined. For example, *brevicornis* types in Lake Winnipeg range above the thermocline into summer-warm waters up to 20°C but breed only in winter when temperatures are near 4°C (Bousfield pers. comm. 1983).

Although *Pontoporeia* species are generally found in deep, cold lakes with well-oxygenated hypolimnions, they can withstand oxygen levels as low as 0.72 cc/l (Juday and Birge 1927). The sternal gills may facilitate respiration and possibly osmoregulation in waters of low oxygen tension.

Diurnal Migration. Vertical migration is not well understood for these amphipods, as it generally involves less than 7% of the population (Mozley and Howmiller 1977). Nocturnal vertical migration of males begins at approximately the same light intensity that stimulates mysid movement, but *Pontoporeia* does not remain at the surface throughout the night as *M. relicta* does (McNaught and Hasler 1966). *Pontoporeia* may ascend 40–70 m in a half hour, but will not readily cross the thermocline (Wells 1960).

Some *Pontoporeia* have been found to remain pelagic during the day, prompting speculation that the vertical migrations may be related to mating activites (Segerstråle 1937; Marzolf 1965a; Wells 1968; Mozley and Howmiller 1977). Movement off the bottom allows for horizontal distribution of the animals by currents and may facilitate mating between otherwise isolated benthic populations.

Food and Feeding Behavior. *Pontoporeia* species are nonselective detritus feeders. In Great Slave Lake they were found to consume detritus, sand grains, algal cells, and diatom frustules (Larkin 1948, Moore 1979b) although they were unable to digest live algae (Kidd 1970).

Laboratory studies showed that *Pontoporeia* abundance could be correlated to bacterial abundance near the surface of the sediment (Marzolf 1965b).

Moore (1979b) found seasonality in the feeding pattern of these amphipods in Great Slave Lake. From May to August, 100% of the organisms captured had food in their guts in contrast to only 10% in December.

As Food for Fish. *Pontoporeia* are a major item in the diet of larger alewife (Morsell and Norden 1968) and smelt (Anderson and Smith 1971). These amphipods are also consumed by cisco (Pearse 1921), burbot (Van Oosten and Deason 1938; Bailey 1972), young lake trout (Dryer et al. 1965), and bloater (Wells and Beeton 1963; Bersamin 1958). Many shallow water species such as carp, bullheads, suckers, trout-perch, spottail shiner, slimy sculpin, and sticklebacks feed heavily on *Pontoporeia* in the summer and fall (Mozley and Howmiller 1977).

Old-squaw ducks have also been found to consume significant numbers of *Pontoporeia* (Mozley and Howmiller 1977).

Unfortunately, *Pontoporeia* may concentrate sublethal levels of toxic materials, pesticides, and radionucleotides from the environment in their tissues. A ten-fold increase in the concentration of DDT and its derivatives was found in alewife, whitefish, and old-squaw ducks feeding on these amphipods (Hickey et al. 1966).

LIFE HISTORY AND ECOLOGY IN OTHER LAKES

In most areas species of *Pontoporeia* appear to have a 2-year life cycle, with mature males living about a week

(Green 1968) and mature females brooding the young for 4 months before releasing them and dying (Segerstråle 1967). However, in the warmer waters of Lake Washington, Washington, the life cycle is condensed to 1 year (Green 1968). In a relatively warm, mesotrophic Russian lake with high food availability, the life cycle of *Pontoporeia* (*affinis*?) was completed in 8 months with the females carrying large broods of 32–34 eggs (Kuzmenko 1969). While warm temperatures reduce the life span, cold temperatures may pro-

long it. Larkin (1948) postulated that *Pontoporeia* in Great Slave Lake may live up to 3 years, and some females could produce a second brood.

In Green Lake, Wisconsin these amphipods were found to carry 13–28 eggs (Juday and Birge 1927). Egg number may be related to the size of the parent. In Great Slave Lake (Moore 1979b), 6-mm females carried 16 eggs, while 9-mm females had 22 eggs. These reports probably represent several species.

Notes on the Distribution and Abundance of the Less Common Crustacean Zooplankton

The following 70 species have been found in at least one of the Great Lakes. They generally occur in low densities in the limnetic zone but may be quite abundant in certain littoral habitats.

ORDER CLADOCERA

Family Sididae

Latona setifera (O. F. Müller) 1785 is a large cladoceran (2.0–3.0 mm) that is distinguished from the more commonly occurring *Diaphanosoma* by the lateral expansion on the basal joint of the dorsal branch of its antennae (Fig. 17) (Brooks 1959; Frey 1960). This species is rare in Lakes Huron (Bigelow 1922), Erie (Wilson 1960; Watson et al. 1974), and Michigan (Stewart 1974; Evans et al. 1980). It was collected from the Duluth-Superior Harbor of Lake Superior during the summer of 1978 (Balcer, unpublished). *L. setifera* is occasionally consumed by trout-perch, freshwater drum, and carpsuckers in Lake Erie (Ewers 1933; Kinney 1950; Wilson 1960).

Latonopsis occidentalis Birge 1891 reaches a length of 1.8 mm (Brooks 1959) and is distinguished from *Diaphanosoma* by its ocellus (Fig. 18). *L. occidentalis* has only been reported from Lake Erie, where it was found in the stomachs of yellow perch (Wilson 1960).

Sida crystallina (O. F. Müller) 1785 has been reported from Lake Superior (Forbes 1891; Selgeby 1974, 1975a; Watson and Wilson 1978), Lake Huron (Sars 1915; Bigelow 1922; Carter 1972), Lake Erie (Chandler 1940; Wilson 1960), Lake Ontario (Tressler et al. 1953), and Lake Michigan (Eddy 1927; Wells 1960; Gannon 1972a, 1974; Stewart 1974; Evans et al. 1980). This species has a patchy distribution, occurring in low numbers in some areas (Selgeby 1975a) but reaching concentrations of 370–5000 organisms m^{-3} in other areas (Eddy 1927; Tressler et al. 1953). *S. crystallina* is a large cladoceran (3–4 mm) with three segments in the dorsal branch of the antennae (Fig. 19) (Brooks 1959; Frey 1960). It is an important food item of cisco (Bigelow 1922) and is also consumed by bass, yellow perch, logperch, crappie, sauger, and freshwater drum in the Great Lakes (Wickliff 1920; Ewers 1933; Wilson 1960). Bottrell (1975), Thomas (1963), and Scourfield (1941) present more details on the ecology of this organism.

Family Daphnidae

Daphnia ambigua Scourfield 1947 has recently been collected in low numbers from Lakes Michigan (Wells 1970), Huron (Carter 1972), and Erie (Patalas 1972). This is a very small species of *Daphnia* (0.75–1.0 mm), which possesses an ocellus and uniformly fine pectens on the postabdominal claw (Fig. 50). More detailed descriptions are provided by Brooks (1957, 1959).

Daphnia parvula Fordyce 1901 has been observed in Lake Michigan's Green Bay and Milwaukee Harbor (Howmiller and Beeton 1971; Gannon 1972a, 1974) and near the D.C. Cook Power Plant in the southeast portion of the lake (Evans et al. 1980). Only one specimen has been collected from the open lake (Torke 1975). *D. parvula* lacks an ocellus and has slightly enlarged teeth in the middle pecten of the postabdominal claw (Fig. 48). The short shell spine and rounded head of this 0.75–1.0 mm species distinguishes it from *D. retrocurva* (Brooks 1957, 1959).

Daphnia pulex Leydig 1860 emend. Richard 1896 has been reported in the zooplankton of Lakes Superior (Smith 1874b; Olson and Odlaug 1966), Erie (Vorce 1881, 1882; Chandler 1940; Langlois 1954; Davis 1954, 1962, 1968; Wilson 1960; Rolan et al. 1973; Britt et al. 1973), and Michigan (Gannon 1974; Evans et al. 1980). It has also been found in the stomachs of fish collected from Lakes Ontario (Clemens and Bigelow 1922) and Erie (Clemens and Bigelow 1922; Sibley 1929; Ewers 1933; Wilson 1960). Olson and Odlaug (1966) and Davis (1954) reported concentrations of *D. pulex* in Lakes Superior and Erie exceeding 5000 m⁻³ and 12000 m⁻³, respectively. It is most likely that these investigators classified all *Daphnia* with an enlarged middle pecten on the postabdominal claw (including *D. retrocurva*) as *D. pulex* (Brooks 1957). Most recent studies (Britt et al. 1972; Rolan et al. 1973; Gannon 1975) show that *D. pulex* as described by Brooks (1957) with its ocellus and short shell spine (Fig. 49), is present in low numbers in Lakes Erie and Michigan. Dodson (1981) and Brandlova et al. (1972) concluded that several species may actually be present in this species complex, which includes animals from 1.3–2.2 mm long.

Daphnia schødleri Sars 1862 reaches a length of 1.5–2.0 mm and possesses a small ocellus and a greatly enlarged middle pecten on the postabdominal claw (Fig. 51). It is generally distinguished from *D. pulex* by its elongate shell spine (Brooks 1957, 1959). Recent studies by Grogg (1977) and Dodson (1981) have shown that *D. schødleri* is morphologically similar to *D. pulicaria* (not reported from the

Great Lakes). *D. schødleri* has recently been collected from Lake Michigan (Wells 1970; Gannon 1972a, 1974).

Scapholeberis aurita (Fischer) 1849 is occasionally collected near the surface in littoral areas of Lakes Erie (Langlois 1954; Wilson 1960) and Michigan (Wells 1970). This 1.0-mm long species is described by Brooks (1959).

Scapholeberis kingi Sars 1903 (Fig. 47) occurs in Lakes Erie (Wilson 1960), Michigan (Gannon 1974; Stewart 1974; Evans et al. 1980), and Superior (Forbes 1891; Birge 1893). The early reports listed this species as *S. mucronata*, a name no longer recognized. Brooks (1959) describes this 0.8–1.0 mm cladoceran that is consumed by largemouth bass, carp, and carpsuckers in the Great Lakes (Sibley 1929; Ewers 1933; Wilson 1960).

SIMOCEPHALUS SCHØDLER 1858

These species (Fig. 46) are described by Frey (1958, 1960) and Brooks (1959).

Simocephalus exspinosus (Koch) 1841, which reaches a length of 3.0 mm, has only been found in Lake Superior (Birge 1893).

Simocephalus serrulatus (Koch) 1841 is present in Lake Michigan (Gannon 1974; Evans et al. 1980) and is abundant in the littoral areas of Lake Erie where it is consumed by bullhead and carp (Wilson 1960).

Simocephalus vetulus Schødleri 1858 is occasionally found in Lake Superior (Birge 1893; Conway et al. 1973) and Lake Michigan (Evans et al. 1980). It is more abundant in the littoral areas of Lake Erie where it serves as food for carp and bullhead (Wilson 1960).

Family Moinidae

Moina micrura Kurz 1874 (Fig. 45) is a relatively small cladoceran, 0.5–0.6 mm long (Brooks 1959), recently collected from Lake Michigan (Wells 1960; Gannon 1972a, 1974). It generally occurs in low numbers but may reach densities of 500–1000 m⁻³ in July in Green Bay (Gannon 1974).

Family Macrothricidae

Acantholeberis curvirostris (O. F. Müller) 1776 (Fig. 38) is a fairly large cladoceran (females to 1.8 mm) that is usually found in shallow, weedy areas and sphagnum bogs (Brooks 1959). It was collected from the shallow waters of Lake Huron's Georgian Bay in 1903 by Bigelow (1922).

Drepanothrix dentata (Eurén) 1861 (Fig. 39) is a rare cladoceran reported from Isle Royale in Lake Superior (Birge 1893) and Georgian Bay in Lake Huron (Bigelow 1922). This animal reaches a length of 0.7 mm and is described by Brooks (1959).

Ilyocryptus acutifrons Sars 1862, which reaches a length of 0.7 mm, has recently been reported from southeastern Lake Michigan (Evans et al. 1980). Brooks (1959) provides a description of this species.

Ilyocryptus sordidus (Liéven) 1848 (Fig. 40) is rare in the zooplankton of Lakes Michigan (Gannon 1972a, 1974; Stewart 1974; Evans et al. 1980) and Erie (Wilson 1960; Rolan et al. 1973). This littoral species, which reaches a length of 1.0 mm, is described by Brooks (1959) and Frey (1960).

Ilyocryptus spinifer Herrick 1884 occurs in Lakes Erie (Wilson 1960; Bradshaw 1964) and Ontario (McNaught and Buzzard 1973). Birge (1893) also reported this species from Lake Superior but used the term *I. longiremus* (Selgeby 1975b). Abundance is quite variable but may exceed 3000 m⁻³ in Lake Erie (Bradshaw 1964). This medium-size cladoceran (0.8 mm) is occasionally eaten by carp and carpsuckers (Wilson 1960). It is described in more detail by Brooks (1959).

Macrothrix laticornis (Jurine) 1820 (Fig. 41) is only 0.5–0.7 mm long (Brooks 1959). It is rarely found in Lakes Erie (Wilson 1960), Michigan (Gannon 1972a; Stewart 1974, Evans et al. 1980), Superior (Birge 1893), and Ontario (McNaught and Buzzard 1973; Czaika 1978b).

Ophryoxus gracilis Sars 1861 (Fig. 42) is a fairly large macrothricid, reaching a length of 2.0 mm (Brooks 1959; Frey 1960). It occurs in the littoral zones of Lake Huron's Georgian Bay (Bigelow 1922) and Lake Michigan's Green Bay (Gannon 1974).

Wlassicsia kinistinensis Birge 1910 (Fig. 43) is a macrothricid occasionally found among the nearshore vegetation in Lake Michigan during the summer (Gannon 1970). It reaches a length of 0.8 mm (Brooks 1959).

Family Chydoridae

Acroperus harpae Baird 1843 (Fig. 22) occurs in the littoral zones of Lakes Michigan (Gannon 1974), Erie (Langlois 1954; Wilson 1960), and Superior (Smith and Moyle 1944; Selgeby 1975a). It is generally present in low numbers (< 1m⁻³) in Lake Superior (Selgeby 1975a) but is quite abundant in some areas of Green Bay (Gannon 1974). Two other organisms, *A. leucocephalus* (Forbes 1887; Birge 1893) and *A. angustatus* (Birge 1893; Wilson 1960) are mentioned in the Great Lakes zooplankton literature. *A. leucocephalus* has been synonymized with *A. harpae*, while *A. angustatus* is merely a crested form of the same species. Frey (1958, 1960) and Brooks (1959) describe this small cladoceran (0.4–0.9 mm long) that is occasionally eaten by Great Lakes fish (Pearse 1921; Sibley 1929). The life history and vertical migration patterns of *A. harpae* are discussed by Scourfield and Harding (1941), Fryer (1968), Lim and Fernando (1978), and Williams and Whiteside (1978).

ALONA BAIRD 1850

Seven species of this genus occur in the Great Lakes (Evans et al. 1980). They are usually more common at the sediment-water interface than in the plankton (Plate 41, Fig. 21).

Alona affinis (Leydig) 1860 has been found in all the Great Lakes (Birge 1893; Bigelow 1922; Wilson 1960; Stewart 1974; Gannon 1974; Beeton and Barker 1974; Watson et al. 1974; Evans and Stewart 1977; Evans et al. 1980). It is considered a benthic form (Fryer 1968; Evans and Stewart 1977) and is usually collected from near the bottom in weedy areas of the littoral zone. *A. affinis* is the largest member of this genus, with females reaching a length of 1.0 mm. Frey (1958) and Brooks (1959) describe the morphology of this cladoceran while Fryer (1968), Bottrell (1975), and Ward (1940) discuss its life history and ecology.

Alona circumfimbriata Megard 1967 is a small cladoceran (0.4 mm) recently found in the littoral zone of Lake Michigan (Gannon 1974). This species is described by Pennak (1978).

Alona costata Sars 1862 is present in low densities in the littoral zone of Lakes Erie (Wilson 1960), Michigan (Gan-

Distribution and Abundance of Less Common Crustacean Species

non 1974; Evans et al. 1980), and Superior (Selgeby 1974, 1975a). In Lake Erie it is consumed by largemouth bass and yellow perch (Ewers 1933; Wilson 1960). The ecology of this species has been studied in several lakes (Ward 1940; Frey 1965; Fryer 1968). Its morphology is described by Brooks (1959) and Frey (1960).

Alona guttata Sars 1862 is a small cladoceran (< 0.4 mm) that occurs in low densities (< 1 m^{-3}) in the littoral zone of Lake Superior (Birge 1893; Conway et al. 1973; Selgeby 1975a) and in Lake Michigan (Evans et al. 1980). Although it has not yet been reported from the plankton of Lake Erie, *A. guttata* has been found in the stomachs of largemouth bass taken from that lake (Ewers 1933). More information on the morphology and ecology of this species is presented by Brooks (1959), Frey (1960, 1965), Ward (1940), Scourfield and Harding (1941), and Fryer (1968).

Alona lepida Birge 1893 was collected as plankton near Isle Royale in Lake Superior. This species is no longer listed as occurring in North America (Brooks 1959; Pennak 1978), and it is uncertain which species Birge actually collected.

Alona quadrangularis (O. F. Müller) 1785 occurs in the littoral zones of Lake Michigan (Evans et al. 1980) and Lake Erie (Langlois 1954), where it serves as a food item for carp and suckers. Frey (1960) and Brooks (1959) describe this cladoceran, which reaches a length of 0.9 mm.

Alona rectangula Sars 1861 grows to a length of 0.5 mm (Fryer 1968). It is found in the littoral region of Lake Michigan (Evans et al. 1980) and Lake Erie (Wilson 1960) and may be consumed by fish feeding in these areas (Sibley 1929). The morphology (Frey 1958, 1960; Brooks 1959) and ecology (Ward 1940; Fryer 1968) of this species have been studied in other lakes.

ALONELLA SARS 1862

The North American members of the genus *Alonella* are described by Frey (1958, 1959, 1960, 1961), Brooks (1959), and Smirnov (1969) (Fig. 14).

Alonella excisa (Fischer) 1854 reaches a length of 0.5 mm. It has only been collected from Lake Erie by Wilson (1960) who also found this species in the stomachs of carpsuckers collected from that lake.

Alonella nana (Baird) 1850 is the smallest member of this genus, reaching a length of only 0.28 mm. Birge (1893) collected it from the Isle Royale region of Lake Superior but reported it as *Pleuroxus nanus* (Selgeby 1975b). It also oc-

curs in the littoral zone of Lake Erie, where it is eaten by carp and bullhead (Wilson 1960).

Alonopsis elongata Sars 1861 (Fig. 23) was collected from Lake Superior by Birge (1893) who used the name *A. latissima*. Flossner (1964), Fryer (1968), Smirnov (1966b), and Brooks (1959) provide more information on the ecology and morphology of this rare cladoceran that reaches a length of 0.8 mm.

Anchistropus minor Birge 1893, which reaches a length of 0.35 mm (Fig. 25), is described by Brooks (1959) and Frey (1960). It has been found in Washington Harbor of Isle Royale in Lake Superior (Birge 1893) and among the weeds of Lake Huron's Georgian Bay (Bigelow 1922) and Lake Michigan's Green Bay (Gannon 1974).

Camptocercus macrurus (O. F. Müller) 1785 is occasionally collected from near the bottom in weedy areas of Lake Erie (Vorce 1882; Langlois 1954). It was previously called *Lynceus macrurus* and is described by Brooks (1959) and Frey (1960).

Camptocercus rectirostris Schødler 1862 (Fig. 26) is widely distributed in the littoral zones of Lakes Michigan (Gannon 1972a, 1974; Evans et al. 1980), Ontario (Czaika 1974a, 1978a), Huron (Bigelow 1922), and Erie (Wilson 1960). It may be quite abundant in certain regions with summer densities reaching 1700 organisms m^{-3} (Czaika 1974a). Frey (1958, 1960) and Brooks (1959) provide descriptions of this cladoceran that is < 1 mm long.

Disparalona acutirostris (Birge) 1878, previously known as *Alonella acutirostris*, is found in low numbers (< 1 m^{-3}) in Lake Superior (Selgeby 1975b). In 1971 Fryer allocated this small cladoceran, 0.22–0.40 mm, to the genus *Disparalona*. It is described by Frey (1961).

Disparalona rostrata (Koch) 1841. In 1893 Birge collected a small cladoceran from Lake Superior and identified it as *Alonella rostrata*. In 1968 Fryer moved these animals formerly known as *Lynceus rostratus* and *Alonella rostrata* to the genus *Disparalona*. Evans et al. (1980) recently found this species in southeast Lake Michigan. A description of this small species (females 0.37–0.59 mm, males 0.38–0.40 mm) is provided by Frey (1961).

Eurycercus lamellatus (O. F. Müller) 1785 (Fig. 27) is well described by Herrick and Turner (1895), Frey (1958, 1960), and Fryer (1963). It is a large cladoceran, reaching a length of 3.0 mm (Brooks 1959). *E. lamellatus* occurs in low numbers (< 18 m^{-3}) in the plankton of Lakes Michigan

(Gannon 1972a, 1974, 1975; Stewart 1974; Evans and Stewart 1977; Evans et al. 1980), Superior (Smith 1874a, Forbes 1981; Birge 1893; Selgeby 1975a), Huron (Bigelow 1922), Ontario (Czaika 1974a), and Erie (Wilson 1960; Bradshaw 1964; Rolan et al. 1973; Britt et al. 1973). This benthic organism is most abundant at the sediment-water interface (Evans et al. 1980) and moves up into the water column at night (Evans and Stewart 1974). It may reach densities of 500 m^{-3} (Stewart 1974). In the Great Lakes, bloater, carpsuckers, and yellow perch prey upon *Eurycercus* (Clemens and Bigelow 1922; Wilson 1960; Wells and Beeton 1963). The feeding behavior and ecology of this cladoceran are described by Fryer (1963) and Frey (1971).

Graptoleberis testudinaria (Fisher) 1851 (Fig. 29) is a small cladoceran (0.5–0.7 mm) found in very low numbers in Lakes Huron (Bigelow 1922), Erie (Langlois 1954), and Michigan (Swain et al. 1970b). It is described by Frey (1958, 1960) and Brooks (1959).

Kurzia latissima (Kurz) 1875 (Fig. 28) reaches a length of 0.6 mm (Brooks 1959; Frey 1960). It is occasionally found in Lakes Huron (Bigelow 1922) and Erie (Langlois 1954).

Leydigia acanthocercoides (Fischer) 1854 was found at five nearshore sampling sites in Lake Erie by Wilson (1960). This 1-mm long cladoceran is described by Brooks (1959) and Frey (1960).

Leydigia leydigi (Schoedler) 1863 has been reported from the Great Lakes as *L. quadrangularis* (Leydig) 1860. Brandlova et al. (1972) discuss their basis for the taxonomic change. This species is rarely found in the plankton of Lakes Huron (Bigelow 1922), Erie (Wilson 1960; Rolan et al. 1973), and Michigan (Stewart 1974; Evans et al. 1980). Sars (1895) and Brooks (1959) provide descriptions of this 0.9 mm zooplankter (Fig. 30).

Monospilus dispar Sars 1862 (Fig. 31) is 0.4–0.5 mm long, lacks a compound eye, but has a well-developed ocellus (Brooks 1959). It is found in low numbers (< 6 m^{-3}) in Lakes Huron (Bigelow 1922), Erie (Wilson 1960; Rolan et al. 1973), and Superior (Birge 1893). Birge incorrectly referred to this species as *M. tenuirostris* (Selgeby 1975b).

PLEUROXUS BAIRD 1843

These species (Fig. 32) are described by Brooks (1959) and Frey (1960).

Pleuroxus aduncus (Jurine) 1820 has only been collected from Cedar Bay of Lake Erie, where it serves as food for carp and carpsuckers (Wilson 1960). Females reach a length of 0.6 mm, while males are only 0.45 mm long.

Pleuroxus denticulatus Birge 1879 is present in the littoral zones of Lakes Michigan (Stewart 1974; Gannon 1974; Evans et al. 1980) and Erie (Langlois 1954; Wilson 1960). Females are 0.5–0.6 mm in length, and males grow to 0.36 mm.

Pleuroxus hastatus Sars 1802 was collected from the open waters of Lake Superior by Birge (1893). Brooks (1959) reports females of 0.6 mm and males of 0.45 mm.

Pleuroxus procurvus Birge 1877 is found in Lakes Michigan (Gannon 1974; Evans et al. 1980), Erie (Ewers 1933; Langlois 1954), and Superior (Birge 1893; Selgeby 1975a). Although Birge originally described this species as *P. procurvus*, he refers to it as *P. procurvatus* in his 1893 report from Lake Superior, but this latter name is not in common usage. It reaches a length of 0.5 mm.

Pleuroxus striatus Schødler 1858 has only been found at four nearshore locations in Lake Erie, where it is eaten by carpsuckers (Wilson 1960). Females grow to 0.8 mm, while males are smaller, 0.6 mm long.

Rhynchotalona falcata (Sars) 1861 (Fig. 33) is a small (< 0.5 mm) cladoceran with a recurved rostrum (Brooks 1959; Frey 1960). It occurs in low densities (< 1 m^{-3}) in Lake Superior (Selgeby 1975a). Birge (1893) also collected this species from Lake Superior but referred to it as *Alona falcata*.

ORDER EUCOPEPODA, SUBORDER CALANOIDA

Family Centropagidae

Osphranticum labronectum S. A. Forbes 1882 was collected in Lake Michigan by Swain et al. (1970b). This species has also been reported from Lake Superior (Swain et al. 1970a; Conway et al. 1973). Robertson and Gannon (1981) question this report, and Selgeby (1975b) suggests that the records from Lake Superior may be based on misidentifications of *Senecella calanoides*. *Osphranticum*, which reaches a length of 1.5 mm, is recognized by its short caudal rami that possess five terminal setae of unequal length.

Family Diaptomidae

Skistodiaptomus reighardi (Marsh) 1895 is occasionally found in Lakes Erie (Jahoda 1948; Rolan et al. 1973), Huron (Carter 1972), and Michigan (Stewart 1974; Evans et al. 1980). This species reaches a length of 1.2 mm and is described by Jahoda (1948), Wilson (1959), and Czaika and Robertson (1968). It resembles *S. oregonensis*, but adult male *S. reighardi* are distinguished by the long angular claw on the exopod of the right fifth leg. Females have two small lateral spines on the genital segment (Figs. 74, 83).

Skistodiaptomus pallidus (Herrick) 1879 occurs in low densities in Lakes Ontario (Patalas 1969; Czaika 1974a, 1978a) and Erie (Patalas 1972; Cap 1979). It is very similar in appearance to *S. oregonensis*. The shape of the fifth legs of males and females is used to distinguish these species (Wilson 1959; Czaika 1974a) (Fig. 71 and 77). Friedman and Strickler (1975) determined that *S. pallidus* is an obligate filter feeder but is capable of rejecting undesirable food particles.

ORDER EUCOPEPODA, SUBORDER CYCLOPOIDA

Family Cyclopidae

Cyclops scutifer Sars 1863 is distinguished from the more common *Diacyclops thomasi* and *Acanthocyclops vernalis* by possessing a row of fine hairs on the inner margins of the caudal rami (Fig. 92). A dorsal longitudinal ridge is also present on each ramus (Yeatman 1959; Torke 1976). Females reach a length of 1.3–1.9 mm, while males are 1.0–1.4 mm long. This species has only been reported from Georgian Bay in Lake Huron (Carter 1972). Its life cycle is described by Paquette and Pinel-Alloul (1982).

Cyclops strenuus Fischer 1851 was collected from Lake Superior by Selgeby (1975a, 1975b) and may have entered this lake via diverted tributaries that once flowed north into James Bay. Yeatman verified Selgeby's identification (Selgeby 1975b), but in a later report Torke (1976) credits Selgeby with finding *C. scutifer* and not *C. strenuus* in Lake Superior. These two species are very similar in form, but *C. strenuus* (Fig. 93) possesses longer caudal rami

(Yeatman 1959; Smith and Fernando 1978). Females are 1.4–2.4 mm long, while males only reach 1.3–1.6 mm.

Diacyclops nanus (Sars) 1863 (Fig. 88) is a benthic cyclopoid that occurs in Lakes Huron's Georgian Bay (Hare and Carter 1976) and Lake Erie (Evanko 1977). Samples from Georgian Bay taken with an Ekman grab show densities of 3300 organisms m^{-2}. This small animal (0.4–0.9 mm) has 11-segmented first antennae and fifth legs with 2 distinct segments (Yeatman 1959; Pennak 1978).

Eucyclops agilis (Koch) 1838 is occasionally referred to as *E. serrulatus* (Torke 1976). It occurs in Lakes Superior (Forbes 1887; Selgeby 1974, 1975a), Michigan (Gannon 1972a, 1974; Stewart 1974; Evans and Stewart 1977; Evans et al. 1980), and Erie (Wilson 1960; Rolan et al. 1973; Watson and Carpenter 1974). This is primarily a benthic cyclopoid (Gannon 1972a; Evans and Stewart 1977; Evans et al. 1980) but may reach densities of 100 m^{-3} in the plankton (Rolan et al. 1973). In Lake Erie *E. agilis* is eaten by bass, bullhead, shiners, redhorse, carp, and stonerollers (Wickliff 1920; Sibley 1929; Ewers 1933; Wilson 1960). This species is recognized by the spine on the outer corner of its moderately short caudal rami (Fig. 91). The first antennae are 12-segmented and may extend to the end of the metasome. The fifth legs of this medium-size cyclopoid (0.6–1.5 mm) have only one distinct segment (Yeatman 1959; Torke 1976; Pennak 1978).

Eucyclops prionophorus Kiefer 1931. This species has recently been collected from southeastern Lake Michigan (Evans et al. 1980). Consult Yeatman (1959) for the identification of this 0.7–0.9 mm animal. It is distinguished from other *Eucyclops* species by possessing short first antennae that do not extend past the first body segment.

Eucyclops speratus (Lilljeborg) 1901 is similar in size and appearance to *E. agilis* but has longer caudal rami (Yeatman 1959; Torke 1976; Pennak 1978) (Fig. 90). It is occasionally found in Lake Superior (Selgeby 1974), Lake Erie (Watson and Carpenter 1974; Watson 1976), and Lake Michigan (Stewart 1974). Females reach a length of 0.75–0.8 mm.

Macrocyclops albidus (Jurine) 1820 is a fairly large cyclopoid (females 1.5–2.5 mm) that is distinguished by the moderately long inner apical seta of the caudal ramus (Fig. 84). The fifth legs are composed of two segments while the first antennae are 17-segmented (Yeatman 1959; Torke 1976; Pennak 1978). This species is found in littoral regions of Lakes Erie (Wilson 1960), Michigan (Gannon 1974), and Superior (Selgeby 1975a). In Lake Erie it serves as food for largemouth bass, freshwater drum, and bullhead

(Wickliff 1920; Ewers 1933; Wilson 1960). Other aspects of the ecology of this carnivorous cyclopoid are discussed by Ewers (1936), Ward (1940), Fryer (1957b), and Hutchinson (1967).

Paracyclops fimbriatus poppei (Rehberg) 1880 is also known as *Cyclops fimbriatus* and *P. poppei* (Torke 1976). It is a benthic species that is occasionally found in the littoral zooplankton of Lakes Erie (Wilson 1960; Rolan et al. 1973), and Michigan (Eddy 1927; Stewart 1974; Evans and Stewart 1977; Evans et al. 1980). This cyclopoid is recognized by its short, 8-segmented first antennae (Fig. 87). It reaches a length of 0.7–0.9 mm and is described in more detail by Yeatman (1959) and Torke (1976).

ORDER EUCOPEPODA, SUBORDER HARPACTICOIDA

This benthic suborder has not been well studied in the Great Lakes. Presently nine species have been identified. Seven of these—*Bryocamptus nivalis* (Wiley 1975), *B. zschokkei*

(Schmeil 1893), *Epactophanes richardi* (Mrazek 1893), *Mesochra alaskana* (M. S. Wilson 1958), *Moraria cristata* (Chappuis 1929), *Nitocra hibernica* (Brady 1880), and *N. spinipes* (Boeck 1864)—occur rarely in samples from Lake Ontario (Czaika 1974a, 1978a). *Bryocamptus nivalis* is also found in Lake Superior (Maschwitz et al. 1976), while *B. zschokkei* has been collected from Lake Michigan (Evans 1975).

Canthocamptus robertcokeri M. S. Wilson 1958 is more widely distributed and has been found in low numbers in Lakes Michigan (Eddy 1927; Gannon 1970, 1972a; Evans et al. 1980), Ontario (Czaika 1974a, 1978a), and Erie (Davis 1962; Rolan et al. 1973; Czaika 1978b). Concentrations of 250 organisms m^{-3} have been found in Cleveland Harbor (Rolan et al. 1973).

Canthocamptus staphylinoides Pearse 1905 is also found in Lakes Michigan (Stewart 1974), Ontario (Czaika 1974a), and Erie (Chandler 1940; Wilson 1960). In September, Britt et al. (1973) found this species encysted in oval-shaped balls encrusted with mud particles on the bottom of Lake Erie. Czaika (1974b) provides aids to separating *C. robertcokeri* from *C. staphylinoides*.

Summary of Zooplankton Collections from the Great Lakes

	Sampling Date[a]		Sampling Objective	
	Year	Month	or Location	References
LAKE ERIE	1880–1881	Dec.–Jan.	water intake	Vorce 1881
	1882	Mar.–July	water intake	Vorce 1882
	1919–1920		fish stomachs	Wickliff 1920
	1919–1920		fish stomachs	Clemens and Bigelow 1922
	1926	summer	zooplankton	Langlois 1954
	1928		fish stomachs	Sibley 1929
	1928–1929	June–Dec.	zooplankton	Wilson 1960
	1928, 1929, 1930		zooplankton	Wright 1955
	1929	summer	fish stomachs	Ewers 1933
	1929, 1930	summer	benthos	Krecker and Lancaster 1933
	1938–1939	year-round	littoral zooplankton	Chandler 1940
	1939–1950		fish stomachs	Kinney 1950
	1946–1948	year-round	*Leptodora kindti*	Andrews 1949
	1946–1948	Mar.–Dec.	diaptomid spp.	Jahoda 1948
	1946–1948	year-round	cyclopoids	Andrews 1953
	1948, 1949, 1959	Oct., summer	zooplankton	Bradshaw 1964
	1950, 1951	year-round	Cleveland Harbor	Davis 1954
	1951	year-round	calanoids	Davis 1961
			zooplankton	Hubschman 1960
	1956–1957	Mar.–Dec.	Cleveland Harbor	Davis 1962
	1957–1958		fish stomachs	Price 1963
	1958	July	copepods	Davis 1959a
	1961	June–July	*Eurytemora affinis*	Engel 1962a, 1962b
	1961	June–Sept.	zooplankton	Britt et al. 1973
	1961, 1962		fish stomachs	Hohn 1966

	Sampling Date[a]		Sampling Objective	
	Year	Month	or Location	References

LAKE ERIE (*continued*)

	Year	Month	or Location	References
	1962		fish stomachs	Wolfert 1965
	1962		*Eurytemora affinis*	Faber and Jermolajev 1966
	1967	July	zooplankton	Davis 1968
	1967, 1968	Oct., Jan.	zooplankton	Davis 1969
	1968	June–Aug.	zooplankton	Patalas 1972
	1968, 1970		zooplankton	Heberger and Reynolds 1977
	1970	Apr.–Dec.	zooplankton	Watson and Carpenter 1974
	1970	Apr.–Dec.	zooplankton	Watson 1976
	1971	Apr.–Nov.	*Mysis relicta*	Carpenter et al. 1974
	1971–1972, 1973	year-round	Cleveland Harbor	Rolan et al. 1973
	1973–1974		Cleveland Harbor	Czaika 1978b
	1974	summer	*Pontoporeia*	Barton and Hynes 1976
	1974, 1975	July	*Daphnia* spp.	Boucherle and Frederick 1976
	1976		*Skistodiaptomus pallidus*	Cap 1979
	1979	June	*Mysis*, benthos	Fitzsimons 1981

LAKE HURON

	Year	Month	or Location	References
	1907	summer	Georgian Bay	Sars 1915
	1903, 1905, 1907		Georgian Bay	Bigelow 1922
	1919–1920		fish stomachs	Clemens and Bigelow 1922
	1952–1956	summer, fall	benthos	Teter 1960
	1954–1955		benthos	Henson 1970
	1955–1956		*Mysis*	Beeton 1960
	1955–1956		*Pontoporeia*	Henson 1970
	1956	Aug.	calanoids	Robertson 1966
	1967–1968	June–Nov.	copepods	Carter 1969
	1968	Aug.	zooplankton	Patalas 1972
	1967, 1968, 1969	summer, fall	zooplankton	Swain et al. 1970b
	1970	May–Oct.	Georgian Bay	Carter 1972
	1971	May–Nov.	zooplankton	Watson and Carpenter 1974
	1971	April–Nov.	*Mysis relicta*	Carpenter et al. 1974
	1974	summer	*Pontoporeia*	Barton and Hynes 1976
	1974	May, July, Oct.	benthos	Hare and Carter 1976
	1974	spring, fall	zooplankton, nearshore	Basch et al. 1980
	1974	April–Dec.	Georgian Bay	Carter and Watson 1977
	1974–1975	April–Nov.	zooplankton grazing	McNaught et al. 1980

LAKE MICHIGAN

	Year	Month	or Location	References
	1870		benthos	Stimpson 1871
	1871		zooplankton	Smith 1874b
	1872		fish stomachs	Smith 1874a
	1881		benthos, pelagic zooplankton	Forbes 1882
	1887–1888	year-round	littoral and surface zooplankton	Eddy 1927
	1893	Aug.	zooplankton	Birge 1893
	1925		fish stomachs	Creaser 1929
	1926, 1927	Oct., May, July	zooplankton	Eddy 1927
	1926–1927		zooplankton	Eddy 1934
	1931–1932		benthos	Eggleton 1936
	1932		fish stomachs	Van Oosten and Deason 1938

Summary of Zooplankton Collections from the Great Lakes

	Sampling Date[a]		Sampling Objective	
	Year	Month	or Location	References
LAKE MICHIGAN (*continued*)	1933–1936		fish stomachs	Schneberger 1936
	1954–1955	May–Nov.	fish stomachs	Bersamin 1958
	1954, 1955, 1958	summer	pelagic zooplankton	Wells 1960
	1954, 1955	summer–fall	*Mysis relicta*	Beeton 1960
	1954–1961		fish stomachs	Wells and Beeton 1963
	1955–1956		*Pontoporeia*	Henson 1970
	1961	summer	*Pontoporeia*	Marzolf 1965b
	1963	Aug.	benthic and pelagic zooplankton	Marzolf 1965a
	1964	May–Nov.	calanoids	Robertson 1966
	1964	Apr.–Oct.	*Pontoporeia*	Wells 1968
	1964, 1965	Aug., May, June	pelagic zooplankton	McNaught and Hassler 1966
	1966, 1967		fish stomachs	Norden 1968
	1966, 1968	June–Aug.	pelagic zooplankton	Wells 1970
	1966–1967	year-round	fish stomachs	Morsell and Norden 1968
	1967		*Pontoporeia*	Powers and Alley 1967
	1967	June	*Mysis relicta*	Robertson et al. 1968
	1967–1969	summer, fall	zooplankton	Swain et al. 1970b
	1969–1970	year-round	pelagic zooplankton	Gannon 1975
	1969–1970	year-round	Green Bay	Gannon 1974
	1969–1970	year-round	Milwaukee and offshore	Gannon 1972a
	1970–1971	year-round	*Mysis relicta*	Reynolds and Degraeve 1972
	1971	July	zooplankton	Howmiller and Beeton 1971
	1972	Apr.–Nov.	nearshore zooplankton	Roth 1973
	1971–1972	Sept.–June	Milwaukee and offshore	Beeton and Barker 1974
	1972	Apr.–Oct.	littoral zooplankton	Stewart 1974
	1972–1977	May–Dec.	nearshore zooplankton	Evans et al. 1980
	1973	Apr.–Nov.	nearshore zooplankton	Roth and Stewart 1973
	1973–1974	year round	pelagic zooplankton	Torke 1975
	1974	July	benthic and epibenthic zooplankton	Evans and Stewart 1977
	1975–1976	year-round	*Mysis relicta*	Morgan and Beeton 1978
	1975–1976	Dec.–Oct.	diaptomid spp.	Bowers 1977
	1975–1977	Apr.–Dec.	nearshore zooplankton	Hawkins and Evans 1979
	1976	Apr.–June	*Mysis relicta*	Grossnickle and Morgan 1979
LAKE ONTARIO	1919–1920		fish stomachs	Clemens and Bigelow 1922
	1927–1928		fish stomachs	Pritchard 1929
	1939–1940	year-round	pelagic zooplankton	Tressler et al. 1953
	1962		*Eurytemora affinis*	Faber and Jermolajev 1966
	1964	Aug.–Sept.	benthos	Johnson and Matheson 1968
	1964	Sept.	calanoids	Robertson 1966
	1966, 1967, 1968	summer	benthos	Johnson and Brinkhurst 1971
	1967, 1968	June–Oct., Sept.	zooplankton	Patalas 1969
	1967	June–Aug.	zooplankton	Patalas 1972
	1969, 1972	fall, summer	zooplankton	McNaught and Buzzard 1973
	1970		*Limnocalanus macrurus*	Roff 1972
	1970	year-round	zooplankton	Watson and Carpenter 1974
	1971	Apr.–Nov.	*Mysis relicta*	Carpenter et al. 1974
	1971–1972	year-round	pelagic zooplankton	Wilson and Roff 1973

	Sampling Date[a]		Sampling Objective	
	Year	Month	or Location	References
LAKE ONTARIO (*continued*)	1972	Apr.–Dec.	zooplankton	Czaika 1974a
	1972, 1973	Apr.–Dec., spring	zooplankton	Czaika 1978a
	1974	summer	*Pontoporeia*	Barton and Hynes 1976
LAKE SUPERIOR	1870		epibenthos	Smith 1871
	1872	Aug.–Sept.	fish stomachs	Smith 1874a
	1871–1872		zooplankton	Smith 1874b
	1889	Aug.	surface zooplankton	Forbes 1891
	1893	Aug.	zooplankton	Birge 1893
	1913		fish stomachs	Hankinson 1914
	1928	July, Aug.	zooplankton	Eddy 1934
	1933, 1934	year-round	zooplankton	Eddy 1943
	1950–1963		fish stomachs	Dryer et al. 1965
	1959–1961		benthos	Thomas 1966
	1960	summer	pelagic zooplankton	Putnam 1963
	1964	Aug.	zooplankton	Olson and Odlaug 1966
	1966–1969		fish stomachs	Bailey 1972
	1967–1969	summer, fall	pelagic zooplankton	Swain et al. 1970a
	1968	Aug.	zooplankton	Patalas 1972
	1970–1971	June–Nov.	zooplankton	Conway et al. 1973
	1971	Apr.–Nov.	*Mysis relicta*	Carpenter et al. 1974
	1971	June–Nov.	*Limnocalanus macrurus*	Conway 1977
	1971	May–Dec.	Apostle Islands	Selgeby 1974
	1971–1972	year-round	Sault Ste. Marie	Selgeby 1975
	1973	May–Nov.	zooplankton	Watson and Wilson 1978
	1974	summer	*Pontoporeia*	Barton and Hynes 1976
	1974	spring, fall	nearshore zooplankton	Basch et al. 1980
	1979–1980	year-round	western end of lake and Sault Ste. Marie	this study

[a]Blanks indicate unknown or ambiguous information.

Glossary

abdomen region of the body posterior to the thorax (Fig. 2b).

abdominal process a flap or fingerlike extension of the abdomen. *Daphnia* have several dorsal unpaired abominal processes (Fig. 1).

antennae paired second antennae

antennules paired first antennae

antepenultimate segment second from the last segment.

anterior near the head

apex tip of a pointed or conelike structure.

benthic pertaining to bottom regions of a lake, especially in nearshore areas.

biramous having two branches; copepod legs with one or more basal segments and two distal branches are biramous.

bivoltine producing two generations each year.

brood chamber space between the thorax and carapace of cladocerans and some malacostracans where eggs are carried (Fig. 2).

caeca pouches of the gut.

carapace the shieldlike fold of exoskeleton that covers the body of cladocerans and some malacostracans. The carapace is an outgrowth of the anterior dorsal exoskeleton.

caudal rami the paired one-segmented projections from the last abdominal segment of copepods; each ramus bears terminal setae (Fig. 2b).

cephalic pertaining to the head.

compound eye an eye consisting of a cluster of pigmented light-sensitive cells (Fig. 1).

coxa basal segment of a thoracic leg; the coxae of amphipods are often modified into enlarged plates that protect the gills.

cyclomorphosis seasonal changes in the body shape of some cladocerans.

distal farthest from the body.

dorsal upper or "back" surface of an animal, the side opposite the legs (Fig. 1).

endopod inner branch of a biramous appendage (Fig. 2d).

ephippial egg a diapausing resting egg of cladocerans.

ephippium a modified portion of the carapace of cladocerans that forms a protective covering for resting eggs.

exopod outer branch of a biramous appendage (Fig. 2d).

exoskeleton cuticle or protein covering of arthropods; in zooplankton, the cuticle is flexible, and the protein matrix contains little or no salts such as calcium phosphate or carbonate.

flagellum a slender whiplike projection from near the base of the first antennae of some amphipods.

fornix a shoulder or platelike extension of the carapace anterior and dorsal to the second antennae of cladocerans.

geniculate bent sharply, like a knee.

genital segment   body segment bearing the male or female reproductive pores.

gnathopod   modified first thoracic leg of some malacostracans that is used for grasping (Fig. 36).

head shield   exoskeleton covering the head of cladocerans.

helmet   a pointed or rounded, sometimes sickleshaped elongation of the head of *Daphnia* (Fig. 1).

instar   a stage of development between successive molts in crustaceans and insects.

lateral   toward the sides, away from the midline.

limnetic   pertaining to the open-water region.

littoral   pertaining to the nearshore region.

mandibles   paired small appendages in or near the mouth of crustaceans that are used for tearing or grinding food.

maxillae   paired second maxillae, feeding appendages located posterior to the first maxillae or maxillules.

maxillipeds   paired mouthparts located posterior to the second maxillae.

maxillules   paired first maxillae, feeding appendages located posterior to the mandibles.

medial   toward the midline.

metasomal wings   posterior lateral projections from the dorsal portion of the last segment of the metasome of some copepods (Fig. 2e).

metasome   central region of the copepod body (Fig. 2a, 2e).

mucro   one of the paired, small spinelike projections at the ventral edge of the posterior margin of the carapace of *Bosmina* (Fig. 36).

multivoltine   producing several generations each year.

ocellus   a single, darkly pigmented, light-sensitive eyespot (Fig. 1).

oostegites   expanded plates attached to the base of the pereiopods or the ventral portion of the thorax of some female malacostracans; the oostegites are used to retain eggs in a ventral brood chamber.

parthenogenesis   a type of asexual reproduction similar to vegetative reproduction; most cladocerans reproduce parthenogenetically, with females producing female offspring.

pecten   a comb, usually composed of a row of small spines.

pelagic   pertaining to the open-water region.

pereiopods   thoracic appendages of malacostracans, used for swimming and walking (Fig. 3).

pleopods   abdominal appendages of malacostracans (Fig. 3a, 3c).

postabdomen   portion of the body anatomically posterior to the anus; *Daphnia* have a large postabdominal claw located at the tip of the postabdomen (Fig. 1).

postabdominal claw   see postabdomen.

posterior   away from the head (Fig. 1).

prehensile   having the ability to grasp.

proximal   nearest the body.

raptorial   having appendages adapted for grasping prey.

rostrum   the noselike projection of the head of some cladocerans, such as *Daphnia* (Fig. 1).

seta   a hair or thornlike projection of the exoskeleton set in a socket in the cuticle. Setae are often equipped with a muscle and a nerve.

spermatophore   a package of sperm that is transferred from the male to the female.

spine   a sharp-pointed, thornlike extension of the exoskeleton; it is not jointed at the base of set in a socket.

spinule   a small spine.

statocyst   an organ of equilibrium comprised of a fluid-filled sac with calcium carbonate granules that press on sensory hairs.

sternum   ventral plate of the exoskeleton of some crustaceans.

subitaneous eggs   eggs that begin development without requiring a resting period.

swimming hairs   elongate setae on the second antennae of cladocerans.

telson   a single, medially flattened plate projecting from the last abdominal segment of some malacostracans (Fig. 3a, 3c).

thorax   region of the body between the head and abdomen (Fig. 2b).

uniramous   an appendage having only one branch.

uropods   paired, flattened ventral-lateral appendages on the last one to three abdominal segments of some malacostracans (Fig. 3).

ventral   lower body surface, the side bearing the legs.

vestigial   a small structure that is more fully developed in another species or developmental stage.

Bibliography of Great Lakes Crustacean Zooplankton

Adamstone, F. B. 1928. Relict amphipods of the genus *Pontoporeia*. *Trans. Amer. Microsc. Soc.* 47(3):366–71.

Allan, J. D. 1973. Competition and the relative abundances of two cladocerans. *Ecology* 54(3):484–498.

Alley, W. P. 1968. *Ecology of the burrowing amphipod Pontoproeia affinis in Lake Michigan*. Univ. Mich. Great Lakes Res. Div. Spec. Rep. 36.

American Fisheries Society. 1980. *Common and scientific names of fishes from the United States and Canada*. 4th ed. Amer. Fish. Soc. Spec. Pub. 12.

Anderson, E. D., and L. L. Smith, Jr., 1971. A synoptic study of food habits of 30 fish species from western Lake Superior. *Univ. Minn. Agr. Exp. Stn. Tech. Bull.* 279.

Anderson, R. F. 1967. Preliminary report on biological observations in northern Lake Michigan utilizing Scuba. In *Studies on the Environment and Eutrophication of Lake Michigan*, J. C. Ayers and D. C. Chandler, pp. 136–137. Univ. Mich. Great Lakes Res. Div. Spec. Rep. 30.

Anderson, R. S. 1970. Predator-prey relationships and predation rates for crustacean zooplankters from some lakes in western Canada. *Can. J. Zool.* 48(6):1229–1240.

Andrews, T. F. 1948. The parthenogenetic reproductive cycle of the cladoceran, *Leptodora kindtii*. *Trans. Amer. Microsc. Soc.* 67(1):54–60.

Andrews, T. F. 1949. The life history, distribution, growth, and abundance of *Leptodora kindtii* (Focke) in western Lake Erie. *Doctoral Dissertations Ohio State Univ.* 57: 5–11 (*Abstr.*).

Andrews, T. F. 1953. Seasonal variations in relative abundance of *Cyclops vernalis*, *Cyclops bicuspidatus*, and *Mesocyclops leuckarti* in western Lake Erie from July, 1946 to May, 1948. *Ohio J. Sci.* 53:91–100.

Armitage, K. B., and J. C. Tash. 1967. The life cycle of *Cyclops bicuspidatus thomasi* S. A. Forbes in Leavinworth County State Lake, Kansas, USA. *Crustaceana* (Leiden) 13:94–102.

Aycock, D. 1942. Influence of temperature on size and form of *Cyclops vernalis* Fischer. *J. Elisha Mitchell Sci. Soc.* 58:84–93.

Bailey, M. M. 1972. Age, growth, reproduction and food of the burbot, *Lota lota* (Linnaeus) in southwestern Lake Superior. *Trans. Amer. Fish. Soc.* 101(4):667–674.

Baird, W. 1845. The natural history of British Entomostraca. *Trans. Berwick. Nat. Club*.

Baird, W. 1857. Notes on the food of some fresh-water fishes, more particularly the Vendace and Trout. *Edinburgh New Phil. J.* 6(n.s.):17–24.

Barton, D. R., and H. B. N. Hynes. 1976. The distribution of Amphipoda and Isopoda on the exposed shores of the Great Lakes. *J. Great Lakes Res.* 2(2):207–214.

Basch, R. E., C. H. Pecor, R. C. Waybrant, and D. E. Kenaga. 1980. *Limnology of Michigan's nearshore waters of Lakes Superior and Huron*. U.S. Environmental Protection Agency, Ecol. Res. Ser. EPA-600/3-80-059.

Beeton, A. M. 1960. The vertical migration of *Mysis relicta* in Lakes Huron and Michigan. *J. Fish. Res. Board Can.* 17:517–539.

Beeton, A. M. 1965. Eutrophication of the St. Lawrence Great Lakes. *Limnol. Oceanogr.* 10:240–254.

Beeton, A. M. and J. M. Barker. 1974. *Investigation of the influence of thermal discharge from a large electric power station on the biology and near shore circulation of Lake Michigan. Part A. Biology.* Univ. of Wisc.-Milwaukee, Center for Great Lakes Studies Spec. Rep. 18.

Beeton, A. M. and J. A. Bowers. 1982. Vertical migration of *Mysis relicta* Lovén. *Hydrobiologia* 93:53–62.

Berguson, S. 1971. A study of the food commonly consumed by the crustacean *Limnocalanus macrurus.* Unpublished data, cited in Conway et al. 1973.

Berrill, M. 1969. The embryonic behavior of the mysid shrimp *Mysis relicta. Can. J. Zool.* 47(6):1217–1221.

Bersamin, S. V. 1958. A preliminary study of the nutritional ecology and food habits of the chubs (*Leucichthys* spp.) and their relation to the ecology of Lake Michigan. *Pap. Mich. Acad. Sci., Arts, Lett.* 43:107–118.

Bhajan, W. R., and H. B. N. Hynes. 1972. Experimental study on the ecology of *Bosmina longirostris* (O. F. Müller). *Crustaceana* (Leiden) 23:133–140.

Bigelow, N. K. 1922. Representative Cladocera of southwestern Ontario. *Univ. Toronto Biol. Ser.* 20:111–125.

Binford, M. W. 1978. Copepoda and Cladocera communities in a river-swamp system. *Int. Ver. Theor. Angew. Limnol. Verh.* 20(4):2524–2530.

Birge, E. A. 1893. Notes on Cladocera, III. Descriptions of new and rare species. *Trans. Wis. Acad. Sci., Arts, Lett.* 9(part 2):275–317.

Birge, E. A. 1897. Plankton studies on Lake Mendota. II: The crustacea of the plankton from July, 1894, to December, 1896. *Trans. Wis. Acad. Sci., Arts Lett.* 11:174–448.

Birge, E. A. 1918. The water fleas (Cladocera). In *Fresh-water Biology,* H. B. Ward and G. C. Whipple eds., pp. 676–741. New York: John Wiley and Sons.

Birge, E. A., and C. Juday, 1908. A summer resting stage in the development of *Cyclops bicuspidatus thomasi. Trans. Wis. Acad. Sci., Arts, Lett.* 16:1–9.

Birge, E. A., and C. Juday. 1922. The upland lakes of Wisconsin: the plankton. I. Its quantity and composition. *Wis. Geol. Nat. Hist. Surv. Bull.* 64:1–22.

Bityukov, E. P. 1960. Ob. ekologi *Limnocalanus grimaldii* (Guerne). *Finskogs Zaliva Zool. Zhur.* 39(12):1783–1789.

Boers, J. J., and J. C. H. Carter. 1978. Instar development rates of *Diaptomus minutus* (Copepoda, Calanoida) in a small lake in Quebec. *Can. J. Zool.* 56(8):1710–1714.

Bottrell, H. H. 1975. Generation time, length of life, instar duration and frequency of moulting and their relationship to temperature in eight species of Cladocera from the River Thames, Reading. *Oecologia* 19(2):129–140.

Boucherle, M. M., and V. R. Fredrick. 1976. *Daphnia* swarms in the harbor at Putin Bay. *Ohio J. Sci.* 76(2):90–91.

Bousfield, E. L. 1973. *Shallow-water gammaridean Amphipoda of New England.* Comstock Publ. Assoc. (Cornell Univ. Press), Ithaca, New York.

Bousfield, E. L. 1982. Malacostraca. From *Synopsis and classification of living organisms,* S. P. Parker, ed., pp. 232–293. McGraw Hill, New York.

Bowers, J. A. 1977. The feeding habits of *Diaptomus ashlandi* and *Diaptomus sicilis* in Lake Michigan and the seasonal distribution of chlorophyll at a near shore station. Master's thesis, University of Wisconsin.

Bowers, J. A. 1980. Feeding habits of *Diaptomus ashlandi* and *Diaptomus sicilis* in Lake Michigan, USA. *Int. Rev. Gesamten Hydrobiol.* 65(2):259–267.

Bowers, J. A., and N. E. Grossnickle. 1978. The herbivorous habits of *Mysis relicta* in Lake Michigan. *Limnol. Oceanogr.* 23(4):767–776.

Bowers, J. A., and H. A. Vanderploeg. 1982. *In situ* predatory behavior of *Mysis relicta* in Lake Michigan. *Hydrobiologia* 93:205–216.

Bowers, J. A., and G. J. Waren. 1977. Predaceous feeding by *Limnocalanus macrurus* and *Diaptomus ashlandi. J. Great Lakes Res.* 3(3):234–237.

Bowman, T. E., and A. Long. 1968. Relict populations of *Drepanopus bungei* and *Limnocalanus macrurus grimaldii* from Ellesmere Island NWT. *Arctic* 21(3):172–180.

Bradley, B. P. 1976. The measurement of temperature tolerance: verification of an index. *Limnol. Oceanogr.* 21(4):596–599.

Bradshaw, A. S. 1964. The crustacean zooplankton picture: Lake Erie 1939–49–59; Cayuga 1910–51–61. *Int. Ver. Theor. Angew. Limnol. Verh.* 15:700–707.

Brandl, Z., and C. H. Fernando. 1974. Feeding of the copepod *Acanthocyclops vernalis* on the cladoceran *Ceriodaphnia reticulata* under laboratory conditions. *Can. J. Zool.* 52(1):99–105.

Brandl, Z., and C. H. Fernando. 1975. Food consumption and utilization in two fresh water cyclopoid copepods (*Mesocyclops edax* and *Cyclops vicinus*). *Int. Rev. Gesamten Hydrobiol.* 60(4):471–494.

Brandl, Z., and C. H. Fernando. 1978. Prey selection by the cyclopoid copepods *Mesocyclops edax* and *Cyclops vicinus. Int. Ver. Theor. Angew. Limnol. Verh.* 20(4):2505–2510.

Brandl, Z., and C. H. Fernando. 1979. The impact of predation by the copepod *Mesocyclops edax* (Forbes) on zooplankton in three lakes in Ontario, Canada. *Can. J. Zool.* 57(4):940–942.

Brandlova, J., Z. Brandl, and C. H. Fernando. 1972. The cladocera of Ontario with remarks on some species and distribution. *Can. J. Zool.* 50:1373–1403.

Britt, N. W., J. T. Addis, and R. Engel. 1973. Limnological studies of the island area of western Lake Erie. *Bull. Ohio Biol. Surv.* 4(3):1–88.

Brooks, J. L. 1946. Cyclomorphosis in *Daphnia.* I. An analysis of *Daphnia retrocurva* and *Daphnia galeata. Ecol. Monogr.* 16(4):409–447.

Brooks, J. L. 1957. The systematics of North American *Daphnia. Memoirs of the Connecticut Acad. of Arts and Sci.* 13:1–180.

Brooks, J. L. 1959. Cladocera. In *Fresh-water Biology,* 2nd ed., W. T. Edmondson, ed., pp. 587–656. John Wiley and Sons, New York.

Burckhardt, G. 1900. Faunistische und systematische Studien über des zooplankton der Grosseren Seen der Schweiz und ihrer Grenzgebiete. *Revue Suisse Zool.* 7:353–713.

Burns, C. W. 1969. Relation between filtering rate, temperature, and body size in four species of *Daphnia. Limnol. Oceanogr.* 14(5):693–700.

Buttorina, L. G. 1968. The reproductive organs of *Polyphemus pediculus*. *Tr. Inst. Biol. Vnutr. Vod Acad. Nauk USSR* 17(20):41–57.

Cap, R. K. 1979. *Diaptomus pallidus* Herrick 1879 (Copepoda, Calanoida): a new record for eastern Lake Erie. *Ohio J. Sci.* 79(1):43–44.

Carl, G. L. 1940. The distribution of some cladocera and free-living copepoda in British Columbia. *Ecol. Monogr.* 10(1):55–110.

Carpenter, G. F., E. L. Mansey, and N. F. H. Watson. 1974. Abundance and life history of *Mysis relicta* in the St. Lawrence Great Lakes. *J. Fish. Res. Board Can.* 31(3):319–325.

Carpenter, K. E. 1931. Variations in *Holopedium* species. *Science* 74:550–557.

Carter, J. C. H. 1969. Life cycles of *Limnocalanus macrurus* and *Senecella calanoides*, and seasonal abundance and vertical distribution of various planktonic copepods in Parry Sound, Georgian Bay. *J. Fish. Res. Board Can.* 26:2543–2560.

Carter, J. C. H. 1972. Distribution and abundance of planktonic Crustacea in Sturgeon Bay and Shawanaga Inlet, Georgian Bay, Ontario. *J. Fish. Res. Board Can.* 29:79–83.

Carter, J. C. H., and N. H. F. Watson. 1977. Seasonal and horizontal distributions of planktonic Crustacea in Georgian Bay and North Channel, 1974. *J. Great Lakes Res.* 3:113–122.

Chace, F. A., J. G. Mackin, L. Hubricht, A. H. Banner, and H. H. Hobbs. 1959. Malacostraca, In *Fresh-water biology*, 2nd ed. W. T. Edmondson, ed., pp. 869–901. John Wiley and Sons, New York.

Chandler, D. C. 1940. Limnological studies of western Lake Erie. I. Plankton and certain physical-chemical data of the Bass Islands Region, from September 1938, to November, 1939. *Ohio J. Sci.* 40(6):291–336.

Clemens, W. A., and N. K. Bigelow. 1922. The food of ciscoes (Leucichthys) in Lake Erie. *Univ. Toronto Stud. Biol. Ser. No. 20* and *Ontario Fish Res. Lab.* 3:39–53.

Coker, R. E. 1933. Influence of temperature on size of fresh-water copepods (*Cyclops*). *Int. Rev. Gesamten Hydrobiol.* 29:406–436.

Coker, R. E. 1934a. Nearly related copepods differentiated physiologically as well as morphologically. *J. Elisha Mitchell Sci. Soc.* 49:264–284.

Coker, R. E. 1934b. Reaction of some freshwater copepods to high temperature—with a note concerning the rate of development in relation to temperature. *J. Elisha Mitchell Sci. Soc.* 50:143–159.

Coker, R. E. 1934c. Influence of temperature on form of the freshwater copepod, *Cyclops vernalis* Fischer. *Int. Rev. Gesamten Hydrobiol.* 301:411–427.

Coker, R. E. 1943. *Mesocyclops edax* (S. A. Forbes), *M. leuckarti* (Claus) and related species in America. *J. Elisha Mitchell Sci. Soc.* 59:181–200.

Cole, G. A. 1953. Notes on copepod encystment. *Ecology* 34(1):208–211.

Cole, G. A. 1955. An ecological study of the microbenthic fauna of two Minnesota lakes. *Amer. Midl. Nat.* 53(1):213–230.

Comita, G. W. 1956. A study of a calanoid copepod population in an arctic lake. *Ecology* 37(3):576–591.

Comita, G. W. 1964. The energy budget of *Diaptomus siciloides* Lilljeborg. *Int. Ver. Theor. Angew. Limnol. Verh.* 15(2):646–653.

Comita, G. W. 1968. Oxygen consumption in *Diaptomus*. *Limnol. Oceangr.* 13(1):51–57.

Comita, G. W. 1972. The seasonal zooplankton cycles, production and transformations of energy in Severson Lake, Minnesota. *Arch. Hydrobiol.* 70(1):14–66.

Comita, G. W., and G. C. Anderson. 1959. The seasonal development of a population of *Diaptomus ashlandi* Marsh, and related phytoplankton cycles in Lake Washington. *Limnol. Oceanogr.* 4(1):37–52.

Comita, G. W., and S. J. McNett. 1976. The post embryonic developmental instars of *Diaptomus oregonensis* Lilljeborg 1889 (Copepoda). *Crustaceana* 30(2):123–163.

Comita, G. W., and D. W. Tommerdahl. 1960. The post embryonic developmental instars of *Diaptomus siciloides* Lilljeborg. *J. Morphol.* 107:297–355.

Confer, J. L. 1971. Intrazooplankton predation by *Mesocyclops edax* at natural prey densities. *Limnol. Oceanogr.* 16:663–666.

Confer, J. L., and J. M. Cooley. 1977. Copepod instar survival and predation by zooplankton. *J. Fish. Res. Board Can.* 34(5):703–706.

Conway, J. B. 1977. Seasonal and depth distribution of *Limnocalanus macrurus* at a site on western Lake Superior. *J. Great Lakes Res.* 3(1–2):15–19.

Conway, J. B., O. R. Ruschmeyer, T. A. Olson, and T. O. Odlaug. 1973. The distribution, composition and biomass of the crustacean zooplankton population in western Lake Superior. *Univ. of Minn. Water Resour. Res. Center Bull.* 63.

Cooley, J. M. 1971. The effect of temperature on the development of resting eggs of *Diaptomus oregonensis* (Lilljeborg) (Copepoda, Calanoida). *Limnol. Oceanogr.* 16(6):921–926.

Cooley, J. M. 1978. The effect of temperature on the development of diapausing and subitaneous eggs in several freshwater copepods. *Crustaceana* 35(1):27–34.

Cooley, J. M., and C. K. Minns. 1978. Prediction of egg development times of freshwater copepods. *J. Fish. Res. Board Can.* 35:1322–1329.

Creaser, C. W. 1929. The smelt in Lake Michigan. *Science* 69:623.

Cummings, K. W., R. R. Costa, R. E. Rowe, G. A. Moshiri, and R. M. Scanlon. 1969. Ecological energetics of a natural population of the predaceous zooplankter *Leptodora kindtii* Focke (Cladocera). *Oikos* 20(2):189–223.

Cunningham, L. 1972. Vertical migrations of *Daphnia* and copepods under the ice. *Limnol. Oceanogr.* 17(2):301–303.

Czaika, S. C. 1974a. Crustacean zooplankton of southwestern Lake Ontario in 1972 during the international field year for the Great Lakes. In *Proc. 17th Conf. Great Lakes Res.*, pp. 1–16. International Association Great Lakes Research.

Czaika, S. C. 1974b. Aids to the identification of the Great Lakes harpacticoids *Canthocamptus robertcokeri* and *Canthocamptus staphylinoides*. In *Proc. 17th Conf. Great Lakes Res.*, pp. 587–588. International Association Great Lakes Research.

Czaika, S. C. 1978a. Crustacean zooplankton of southwestern

Lake Ontario in spring 1973 and at the Niagara and Genesee River mouth areas in 1972 and spring 1973. *J. Great Lakes Res.* 4(1):1–9.

Czaika, S. C. 1978b. Crustacean zooplankton in Lake Erie off Cleveland Harbor. *Ohio J. Sci.* 78(1):18–25.

Czaika, S. C. 1982. Identification of nauplii NI–NVI and copepoids CI–CVI of the Great Lakes Calanoid and Cyclopoid Copepods (Calanoida, Cyclopoida, Copepoda). *J. Great Lakes Res.* 8(3):439–469.

Czaika, S. C., and A. Robertson. 1968. Identification of the copepods of the Great Lakes species of *Diaptomus* (Calanoida, Copepoda). In *Proc. 11th Conf. Great Lakes Res.*, pp. 39–60. International Association Great Lakes Research.

Dadswell, M. J. 1974. Distribution, ecology and postglacial dispersal of certain crustaceans and fishes in eastern North America. *Natl. Mus. Nat. Sci. (Ottawa) Publ. Zool.* 11:1–100.

Davis, C. C. 1954. A preliminary study of industrial pollution in the Cleveland Harbor area, Ohio. III. The zooplankton and general ecological considerations of phytoplankton and zooplankton production. *Ohio J. Sci.* 54(6):388–408.

Davis, C. C. 1959a. Osmotic hatching in the eggs of some freshwater copepods. *Biol. Bull.* (Woods Hole) 116:15–29.

Davis, C. C. 1959b. Damage to fish fry by cyclopoid copepods. *Ohio J. Sci.* 59(2):101–102.

Davis, C. C. 1961. Breeding of calanoid copepods in Lake Erie. *Int. Ver. Theor. Angew. Limnol. Verh.* 14:933–942.

Davis, C. C. 1962. The plankton of the Cleveland Harbor area of Lake Erie in 1956–1957. *Ecol. Monogr.* 32:209–247.

Davis, C. C. 1968. The July 1967 zooplankton of Lake Erie. In *Proc. 11th Conf. Great Lakes Res.*, pp. 61–75. International Association Great Lakes Research.

Davis, C. C. 1969. Seasonal distribution, constitution, and abundance of zooplankton in Lake Erie. *J. Fish. Res. Board Can.* 26(9):2459–2476.

Davis, C. C. 1972. Plankton succession in a Newfoundland lake. *Int. Rev. Gesamten. Hydrobiol.* 57(3):367–395.

DeCosta, J., and A. Janicki. 1978. Population dynamics and age structure of *Bosmina longirostris* in an acid water impoundment. *Int. Ver. Theor. Angew. Limnol. Verh.* 20(4):2479–2483.

Deevey, G. B. 1948. The zooplankton of Tisbury Great Pond. *Bull. Bingham Oceanogr. Collect. Yale Univ.* 12(1):1–44.

Deevey, E. S., and G. B. Deevey. 1971. The American species of *Eubosmina* Seligo (Crustacea, Cladocera). *Limnol. Oceanogr.* 16(2):201–218.

Degraeve, G. M., and J. B. Reynolds. 1975. Feeding behavior and temperature and light tolerance of *Mysis relicta* in the laboratory. *Trans. Amer. Fish. Soc.* 104(2):394–397.

De Lent, G. M. 1922. Untersuchungen über Plankton-Copepoden in Niederländischen Gewässern. *Int. Rev. Gesamten Hydrobiol.* 10:76–90.

De Pauw, N. 1973. On the distribution of *Eurytemora affinis* (Poppe) in the western Scheldt Estuary (Copepoda). *Int. Ver. Theor. Angew. Limnol. Verh.* 18(3):1462–1472.

Dodson, S. I. 1981. Morphological variation of *Daphnia pulex*

Leydig (Crustacea, Cladocera) and related species from North America. *Hydrobiologia* 83:101–114.

Dryer, W. R., L. F. Erkkila, and C. L. Tetzloff. 1965. Food of lake trout in Lake Superior. *Trans. Amer. Fish. Soc.* 94:169–176.

Dussart, B. 1969. *Les Copépodes des eaux Continales d'Europe Occidentale. Tome 2: Cyclopoides et biologie.* Editions N. Boubée and Cie, Paris.

Eddy, S. 1927. The plankton of Lake Michigan, *Ill. Nat. Hist. Surv. Bull.* 17(4):203–232.

Eddy, S. 1934. A study of fresh-water plankton communities. *Ill. Biol. Monogr.* 12(4):1–90.

Eddy, S. 1943. Limnological notes on Lake Superior. *Proc. Minn. Acad. Sci.* 11:34–39.

Edmonson, W. T. (ed.). 1959a. *Fresh-water biology*, 2nd ed. John Wiley and Sons, New York.

Edmondson, W. T. 1959b. Methods and equipment. In *Fresh water Biology* 2nd ed., W. T. Edmondson, ed., pp. 1194–1202. John Wiley and Sons, New York.

Edmondson, W. T. and G. G. Winberg, eds. 1971. *IBH Handbook No. 17: A manual on methods for the assessment of secondary productivity in fresh waters.* Blackwell Scientific Publications, Oxford.

Eggleton, F. E. 1936. The deep-water bottom fauna of Lake Michigan. *Mich. Acad.* 21:599–612.

Elgmork, K. 1973. Bottom resting stages of planktonic cyclopoid copepods in meromictic lakes. *Int. Ver. Theor. Angew. Limnol. Verh.* 18(3):1474–1478.

Engel, R. E. 1962a. *Eurytemora affinis*, a calanoid copepod new to Lake Erie. *Ohio J. Sci.* 62:252.

Engel, R. E. 1962b. A quantitative study of the zooplankton in the island region of western Lake Erie. Master's thesis, Ohio State University.

Engel, S. 1976. Food habits and prey selection of Coho Salmon (*Oncorhynchus kisutch*) and Cisco (*Coregonus artedii*) in relation to zooplankton dynamics in Pallette Lake, Wisconsin. *Trans. Amer. Fish. Soc.* 105(5):607–614.

Evanko, M. 1977. *Cyclops nanus* Sars 1863 (Copepoda, Cyclopoida): a new record for Lake Erie. *Ohio J. Sci.* 77:100.

Evans, M. S. 1975. *The 1974 pre-operational zooplankton investigations relative to the Donald C. Cook Nuclear Power Plant.* Univ. Mich. Great Lakes Res. Div. Spec. Rep. 58.

Evans, M. S., R. W. Bothelt, and C. P. Rice. 1982. PCB's and other toxicants in *Mysis relicta*. *Hydrobiologia* 93:205–216.

Evans, M. S., B. E. Hawkins, and D. W. Sell. 1980. Seasonal features of zooplankton assemblages in the nearshore area of southeastern Lake Michigan. *J. Great Lakes Res.* 6(4):275–289.

Evans, M. S., L. M. Sicko-Goad, and M. Omair. 1979. Seasonal occurrence of *Tokophyra quadripartita* (Suctoria) as epibionts on adult *Limmocalanus macrurus* in southeastern Lake Michigan. *Trans. Amer. Microsc. Soc.* 98:102–109.

Evans, M. S., and J. A. Stewart. 1977. Epibenthic and benthic microcrustaceans (copepods, cladocerans, ostracods) from a nearshore area in southeastern Lake Michigan. *Limnol. Oceanogr.* 22(6):1059–1066.

Ewers, L. A. 1930. The larval development of freshwater Copepods. *The Ohio State University Franz Theodore Stone Laboratory Contrib.* No. 3:1–43.

Ewers, L. A. 1933. Summary report of Crustacea used as food by the fishes of the western end of Lake Erie. *Trans. Amer. Fish. Soc.* 63:379–390.

Ewers, L. A. 1936. Propagation and rate of reproduction of some freshwater Copepoda. *Trans. Amer. Microsc. Soc.* 55:230–238.

Faber, D. J., and E. G. Jermolajev. 1966. A new copepod genus in the plankton of the Great Lakes. *Limnol. Oceanogr.* 11:301–303.

Fabian, M. W. 1960. Mortality of fresh water and tropical fish fry by cyclopoid copepods. *Ohio J. Sci.* 60(5):268–270.

Fitzsimons, J. 1981. Occurrence of *Mysis relicta* Loven in the western basin of Lake Erie. *Crustaceana* 41(2):217–218.

Flossner, D. 1964. Zur Cladocerenfauna des Stechlinsee— Gevietes II. *Limnologica* 2:35–103.

Focke, 1844. *Polyphemus kindti. Weserzeitung* Bremen, Sept. 22, 1844.

Forbes, E. B. 1897. A contribution to the knowledge of North American fresh water Cyclopidae. *Bull. Ill. Nat. Hist. Surv.* 5:27–82.

Forbes, S. A. 1882. On some Entomostraca of Lake Michigan and adjacent waters. *Amer. Nat.* 16(7):537–542, 640–649.

Forbes, S. A. 1883. The food of the smaller fresh-water fishes. *Bull. Ill. Nat. Hist. Surv.* 1(6):65–95.

Forbes, S. A. 1885. On the food relations of fresh-water fishes: a summary and discussion. *Bull. Ill. Nat. Hist. Surv.* 2:475–538.

Forbes, S. A. 1888. Notes on the first food of the whitefish. *Trans. Amer. Fish. Soc.* 17:59–66.

Forbes, S. A. 1891. On some Lake Superior Entomostraca. *Rept. U.S.* Comm. *Fish and Fisheries 1887*:701–718.

Fordyce, C. 1900. The Cladocera of Nebraska. *Trans. Amer. Microsc. Soc.* 22:119–174.

Forest, H. E. 1879. On the anatomy of *Leptodora hyalina. J. Microsc. (Oxf)* 2(2):825–834.

Franke, H. 1925. Der Fangapparat von *Chydorus sphaericus. Z. Wiss. Zool.* 125:271–298.

Freidenfelt, T. 1913. Zur Biologie von *Daphnia longiremis* G. O. Sars und *Daphnia cristata* G. O. Sars. *Int. Rev. Hydrobiol.* 6:230–242.

Frey, D. G. 1958. The late-glacial Cladoceran fauna of a small lake. *Arch. Hydrobiol.* 54:209–275.

Frey, D. G. 1959. The taxonomic and phylogenetic significance of the Chydoridae. *Int. Rev. Gesamten Hydrobiol.* 44:27–50.

Frey, D. G. 1960. Remains of animals in Quaternary lake and bog sediments and their interpretation. *Arch. Hydrobiol. Suppl. Ergebnisse der Limnologie* 2:i–ii, 1–114.

Frey, D. G. 1961. Differentiation of *Alonella acutirostris* (Birge) 1879 and *Alonella rostrata* (Koch 1841) (Cladocera, Chydoridae). *Trans. Amer. Microsc. Soc.* 80(2):129–140.

Frey, D. G. 1962. Supplement to: The taxonomic and phylogenetic significance of the head pores of the Chydoridae (Cladocera). *Int. Rev. Gesamten Hydrobiol.* 47(4):603–609.

Frey, D. G. 1965. Differentiation of *Alona costata* Sars from

two related species (Cladocera, Chydoridae). *Crustaceana* (Leiden) 8:159–173.

Frey, D. G. 1971. Worldwide distribution and ecology of *Eurycercus* and *Saycia* (Cladocera). *Limnol. Oceanogr.* 16:254–308.

Frey, D. G. 1980. On the plurality of *Chydorus sphaericus* (O. F. Müller) (Cladocera, Chydoridae), and designation of a neotype from Sjaelsø, Denmark. *Hydrobiologia* 69(1–2):83–123.

Frey, D. G. 1982. G. O. Sars and the Norwegian Cladocera: a continuing frustration. *Hydrobiologra* 96:267–293.

Friedman, M. M., and J. R. Strikler. 1975. Chemoreceptors and feeding in calanoid copepods (Arthropoda, Crustacea). *Proc. Nat. Acad. Sci.* 72(10):4185–4188.

Fryer, G. 1954. Contributions to our knowledge of the biology and systematics of the freshwater copepods. *Schweiz. Z. Hydrol.* 16(1):64–77.

Fryer, G. 1957a. The feeding mechanism of some freshwater cyclopoid copepods. *J. Zool.* (Lond.) 129(1):1–25.

Fryer, G. 1957b. The food of some freshwater cyclopoid copepods—and its ecological significance. *J. Anim. Ecol.* 26:263–286.

Fryer, G. 1963. The functional morphology and feeding mechanisms of the chydorid cladoceran *Eurycercus lamellatus. Trans. R. Soc. Edinburgh* 65(14):335–381.

Fryer, G. 1968. Evolution and adaptive radiation in the Chydoridae (Crustacea; Cladocera): a study in comparative functional morphology and ecology. *Philos. Trans. R. Soc. Lond. B. Biol. Sci.* 254:221–385.

Fryer, G. 1971. Allocation of *Alonella acutirostris* (Birge) (Cladocera, Chydoridae) to the genus *Disparalona. Crustaceana* (Leiden) 21:221–222.

Furst, M. 1981. Results of introductions of new fish food organisms into Swedish Lakes. *Rept. Inst. Freshw. Res. Drottingholm* 59:33–47.

Gannon, J. E. 1970. *An artificial key to the common zooplankton crustacea of Lake Michigan exclusive of Green Bay.* Biotest Laboratories Inc., Northbrook, Illinois.

Gannon, J. E. 1972a. A contribution to the ecology of zooplankton crustacea of Lake Michigan and Green Bay. Ph.D. thesis, University of Wisconsin.

Gannon, J. E. 1972b. Effects of eutrophication and fish predation on recent changes in zooplankton crustacea species composition in Lake Michigan. *Trans. Amer. Microsc. Soc.* 91:82–84.

Gannon, J. E. 1974. The crustacean zooplankton of Green Bay, Lake Michigan. In *Proc. 17th Conf. Great Lakes Res.,* pp. 28–51. International Association Great Lakes Research.

Gannon, J. E. 1975. Horizontal distribution of crustacean zooplankton along a cross-lake transect in Lake Michigan. *J. Great Lakes Res.* 1(1):79–91.

Gannon, J. E., and A. M. Beeton. 1971. The decline of the large zooplankter *Limnocalanus macrurus* Sars (Copepoda, Calanoida) in Lake Erie. In *Proc. 14th Conf. Great Lakes Res.,* pp. 27–38. International Association Great Lakes Research.

Gannon, J. E., and R. S. Stemberger. 1978. Zooplankton as

indicators of water quality. *Trans. Amer. Microsc. Soc.* 97:16–35.

Gerschler, M. W. 1911. Monographie der *Leptodora kindtii*. *Arch. Hydrobiol.* 6:415–466.

Gliwicz, Z. M. 1969. Studies on the feeding of pelagic zooplankton in lakes of varying trophy. *Ekol. Pol. Ser. A* 17:663–708.

Gophen, M. 1976. Temperature effect on lifespan, metabolism and development time of *Mesocyclops leuckarti* (Claus). *Oecologia* (Berl.) 25(3):271–277.

Gophen, M. 1979. Bathymetrical distribution and diurnal migrations of zooplankton in Lake Kinneret (Israel) with emphasis on *Mesocyclops leuckarti* (Claus). *Hydrobiologia* 64(3):199–208.

Goulden, C. E., and D. G. Frey. 1963. The occurrence and significance of lateral head pores in the genus *Bosmina* (Cladocera). *Int. Rev. Gesamten Hydrobiol.* 49:513–522.

Green, R. H. 1968. A summer breeding population of the relict amphipod *Pontoporeia affinis* Lindstrom. *Oikos* 19(2):191–197.

Grogg, C. V. 1977. Morphological variation in the "pulex" group of the genus *Daphnia*. Master's report, Ball State University, Muncie, Indiana.

Grossnickle, N. E. 1978. The herbivorous and predaceous habits of *Mysis relicta* in Lake Michigan. Ph.D. thesis, University of Wisconsin-Madison.

Grossnickle, N. E. 1982. Feeding habits of *Mysis relicta*—an overview. *Hydrobiologia* 93:101–108.

Grossnickle, N. E., and M. D. Morgan. 1979. Density estimates of *Mysis relicta* in Lake Michigan. *J. Fish. Res. Board Can.* 36(6):694–698.

Gurney, R. 1923. The crustacean plankton of the English Lake District. *Zool. J. Linn. Soc.* 35:411–447.

Gurney, R. 1931. *British Freshwater Copepoda Vol. I.* Calanoida Dulau and Co. Ltd., London.

Gurney, R. 1933. *British Freshwater Copepoda Vol. III.* Cyclopoida London R. Soc., London.

Hall, D. J. 1964. A experimental approach to the dynamics of a natural population of *Daphnia galeata mendota*. *Ecology* 45(1):94–112.

Hamilton, J. D. 1958. On the biology of *Holopedium gibberum* Zaddock (Crustacea, Cladocera). *Int. Ver. Theor. Angew. Limnol. Verh.* 13(2):785–788.

Haney, J. F., and D. J. Hall. 1975. Diel vertical migration and filter-feeding activities of *Daphnia*. *Arch. Hydrobiol.* 75(4):413–441.

Hankinson, T. L. 1914. Young whitefish in Lake Superior. *Science* 40(n.s.):239–240.

Hare, L. and J. C. H. Carter. 1976. *Diacyclops nanus* (Cyclopoida, Copepoda) A new record from the St. Lawrence Great Lakes. *J. Great Lakes Res.* 2(2):294–295.

Hasler, A. D. 1937. The physiology of digestion in plankton Crustacea. II. Further studies on the digestive enzymes of (A) *Daphnia* and *Polyphemus*; (B) *Diaptomus* and *Calanus*. *Biol. Bull.* 72:290–298.

Hawkins, B. E., and M. S. Evans. 1979. Seasonal cycles of zoo-

plankton biomass in southeastern Lake Michigan. *J. Great Lakes Res.* 5(3–4):256–263.

Heberger, R. F., and J. B. Reynolds. 1977. *Abundance, composition and distribution of crustacean zooplankton in relation to hypolimnetic oxygen depletion in west-central Lake Erie.* U.S. Fish and Wildl. Serv. Tech. Pap. 93.

Hensey, D., and K. G. Porter. 1977. The effect of ambient oxygen concentration on filtering and respiration rates of *Daphnia galeata mendotae* and *Daphnia magna*. *Limnol. Oceanogr.* 22(5):839–845.

Henson, E. B. 1970. *Pontoporeia affinis* (Crustacea, Amphipoda) in the straits of Mackinac Region. In *Proc. 13th Conf. Great Lakes Res.*, pp. 601–610. International Association Great Lakes Research.

Herrick, C. L. 1884. A final report on the Crustacea of Minnesota included in the orders Cladocera and Copepoda. In *12th An. Rep. Minn. Geol. Nat. Hist. Surv.*, pp. 1–194.

Herrick, C. L. 1887. Contribution to the fauna of the Gulf of Mexico and the South. *Mem. Denison Sci. Assoc.* 1(1):56.

Herrick, C. L., and C. H. Turner. 1895. Synopsis of the Entomostraca of Minnesota, pp 1–337. In *2nd Report of State Zoologist, Minn. Geol. Nat. Hist. Survey.* Pioneer Press Co., St. Paul, Minn.

Hickey, J. J., J. A. Keith, and F. B. Coon. 1966. An exploration of pesticides in a Lake Michigan ecosystem. *J. Appl. Ecol.* 3(suppl.):141–154.

Hohn, M. H. 1966. Analysis of plankton ingested by *Stizostedium vitreum vitreum* (Mitchell) fry and concurrent vertical plankton tows from southwestern Lake Erie, May 1961 and May 1962. *Ohio J. Sci.* 66:193–197.

Holmquist, C. 1963. Some notes on *Mysis relicta* and its relatives in northern Alaska. *Artic* 16:109–128.

Holmquist, C. 1970. The genus *Limnocalanus*. *Z. Zool. Syst. Evolutions Forsch* 8:172–196.

Holmquist, C. 1972. Das Zooplankton der Binnengewässer. V. Mysidacea. *Binnengewässer* 26(1):247–256.

Holway, P. H., and A. W. Cocking. 1953. Observations on the vertical distribution of *Bosmina* in Mulham Tarn in Grasmere. *Ann. Mag. Nat. Hist.* 12(6):717–720.

Howmiller, R. P., and A. M. Beeton. 1971. *Report on a cruise of the R/V Neeskay in central Lake Michigan and Green Bay, 8–14 July 1971.* Univ. of Wis.-Milwaukee Center for Great Lakes Studies Spec. Rep. 13.

Hoy, P. R. 1872. Deepwater fauna of Lake Michigan. *Trans. Wis. Acad. Sci., Arts, Lett. (1870–1872)* 1:98–101.

Hubschman, J. H. 1960. Relative daily abundance of planktonic Crustacea in the island region of western Lake Erie. *Ohio J. Sci.* 60(6):335–340.

Humes, A. H. 1955. The postembryonic developmental stages of a freshwater calanoid copepod, *Epischura massachusettsensis* Pearse. *J. Morphol.* 96:441–472.

Hunt, G. W., and A. Robertson. 1977. The effect of temperature on reproduction of *Cyclops vernalis* Fischer. *Crustaceana* (Leiden) 32(2):169–177.

Husmann, S., H. U. Jacobi, M. P. D. Meijering, and B. Reise. 1978. Distribution and ecology of Svalbard's Cladocera.

Int. Ver. Theor. Angew. Limnol. Vehr. 20(4):2452–2456.

Hutchinson, G. E. 1951. Copepodology for the ornithologist. *Ecology* 32:571–577.

Hutchinson, G. E. 1967. *A treatise on limnology Vol II.* John Wiley and Sons, New York.

International Congress of Zoology (XV, 1958). 1964. International code of zoological nomenclature. Revised Edition. International Trust for zoological nomenclature, London.

Ischregt, G. 1933. Über *Polyphemus pediculus. Arch. Hydrobiol.* 25:261–290.

Jahoda, W. J. 1948. Seasonal differences in distribution of *Diaptomus* (Copepoda) in western Lake Erie. Ph.D. dissertation, Ohio State University.

Janicki, A. J., and J. DeCosta. 1977. The effect of temperature and age structure on P/B for *Bosmina longirostris* in a small impoundment. *Hydrobiologia* 56(1):11–16.

Johnson, D. L. 1972. Zooplankton dynamics in Indiana waters of Lake Michigan in 1970. Master's thesis, Ball State University, Muncie, Indiana.

Johnson, L. 1964. Marine-glacial relicts of the Canadian Arctic islands. *Syst. Zool.* 13:76–91.

Johnson, M. G., and R. D. Brinkhurst. 1971. Associations and species diversity of benthic macroinvertebrates of Bay of Quinte and Lake Ontario. *J. Fish Res. Board Can.* 28(11):1683–1697.

Johnson, M. J., and D. H. Matheson. 1968. Macroinvertebrate communities of the sediments of Hamilton Bay and adjacent Lake Ontario. *Limnol. Oceanogr.* 13(1):99–111.

Juday, C. 1904. The diurnal movement of plankton Crustacea. *Trans. Wis. Acad. Sci., Arts, Lett.* 14:534–568.

Juday, C. 1925. *Senecella calanoides*, a recently described freshwater copepod. *Proc. U.S. Natl. Mus.* 66(4):1–16.

Juday, C., and E. A. Birge. 1927. *Pontoporeia* and *Mysis* in Wisconsin lakes. *Ecology* 8(4):445–452.

Jurine, L. 1820. *Histoire des monocles qui se trouvent aux environs de Genève.* Genève and Paris.

Karabin, A. 1978. The pressure of pelagic predators of the genus *Mesocyclops* (Copepods, Crustacea) on small zooplankton. *Ekol. Pol. Ser. A* 26(20):241–258.

Katona, S. K. 1971. The developmental stages of *Eurytemora affinis* (Poppe 1880) (Copepoda, Calanoida) raised in laboratory cultures including a comparison with the larvae of *Eurytemora americana* (Williams 1906) and *Eurytemora herdmani* (Thompson and Scott 1897). *Crustaceana* 21:5–20.

Kerfoot, W. C. 1974. Egg-size cycle of a cladoceran. *Ecology* 55(6):1259–1270.

Kerfoot, W. C. 1977. Competition in cladoceran communities: the cost of evolving defenses against copepod predation. *Ecology* 58(2):303–313.

Kerfoot, W. C. 1978. Combat between predatory copepods and their prey: *Cyclops, Epischura* and *Bosmina. Liimnol. Oceanogr.* 23(6):1089–1102.

Kerfoot, W. C. 1978. Combat between predatory copepods and their prey: *Cyclops, Epischura* and *Bosmina. Limnol. Oceanogr.* 23(6):1089–1102.

Kibby, H. V., and Rigler, F. H. 1973. Filtering rates of *Limnocalanus. Int. Ver. Theor. Angew. Limnol. Verh.* 18(3):1457–1461.

Kidd, C. C. 1970. *Pontoporeia affinis* (Crustacea, Amphipoda) as a monitor of radioactivity in Lake Michigan. Ph.D. thesis, University of Michigan, Ann Arbor.

Kiefer, F. 1927. Versuch eines systems der Cyclopiden. *Zool. Anz.* 73:302–308.

Kiefer, F. 1929a. Crustacea Copepoda. II. Cyclopoida Gnathostoma. *Das Tierreich* 53:1–102.

Kiefer, F. 1929b. Zur kenntnis der Cyclopoiden Nordamerikas. *Zool. Anz.* 80:305–309.

Kiefer, F. 1932. Versuch eines systems der Diaptomiden (Copepoda, Calanoida). *Zool. Jahrb. Abt. Syst. Oekol. Geogr. Tiere* 63: 451–520.

Kiefer, F. 1938. Ruderfuskrebse (Crustacea, Copepoda) aus Mexiko. *Zool. Anz.* 123:274–280.

Kiefer, F. 1960. Ruderfuskrebse (Copepoda). Kosmos-Verlag, Stuttgart.

Kiefer, F. 1978a. Freilebende Copepoda. *Binnengewässer* 26(2):1–343.

Kiefer, F. 1978b. Zur kenntnis des *Diacyclops thomasi* (S. A. Forbes 1882). *Crustaceana* (Leiden) 34(2):214–216.

Kincaid, T. 1953. *A contribution to the taxonomy and distribution of the American fresh-water Calanoid Crustacea.* Calliostoma Co., Seattle, Washington.

Kinney, E. C. 1950. The life history of the trout perch, *Percopsis omiscomaycus* (Walbaum) in western Lake Erie. Master's thesis, Ohio State University.

Kinsten, B., and P. Olsén. 1981. Impact of *Mysis relicta* Lovén introduction on the plankton of two mountain lakes, Sweden. *Rept. Inst. Freshw. Res. Drottingholm* 59:64–77.

Kiser, R. W. 1950. *A revision of the North American species of the cladoceran genus Daphnia.* Edward Bros., Ann Arbor, Michigan.

Kořínek, V. 1981. *Diaphanosoma birgei*—new species (Crustacea, Cladocera) from America and its widely distributed subspecies *Diaphanosma birgei lacustris* new subspecies. *Can. J. Zool.* 59(6):1115–1121.

Krecker, F. H. and L. Y. Lancaster. 1933. Bottom shore fauna of western Lake Erie: a population study to a depth of six feet. *Ecology* 14(2):79–93.

Kuzmenko, K. N. 1969. The life cycle and production of *Pontoporeia affinis* Lindstrom in Lake Krasnoye (Karelian Isthmus). *Hydrobiol. J.* 5(4):53.

Lai, H. C., and J. C. H. Carter. 1970. Life cycle of *Diaptomus oregonensis* Lilljeborg in Sunfish Lake, Ontario. *Can J. Zool.* 48:1299–1302.

Lane, P. A. 1978. Role of invertebrate predation in structuring zooplankton communities. *Int. Ver. Theor. Angew. Limnol. Verh.* 20(1):480–485.

Langeland, A. 1978. Effect of fish (*Salvelinus alpinus*, arctic char) predation on the zooplankton in the Norwegian Lakes. *Int. Ver. Theor. Angew. Limnol. Verh.* 20(3):2065–2069.

Langford, R. R. 1938. Diurnal and seasonal changes in the distribution of the limnetic Crustacea of Lake Nipissing, Ontario.

Univ. Toronto. Stud. Biol., *No. 45* and *Ontario Fish. Res. Lab.* 56:1–142.

Langlois, T. H. 1954. The western end of Lake Erie and its ecology. J. W. Edwards, Ann Arbor, Michigan.

Larkin, P. A. 1948. *Pontoporeia* and *Mysis* in Athabaska, Great Bear, and Great Slave Lakes. *Bull. Fish. Res. Board. Can.* 78:1–33.

Lasenby, D. C., and M. Fürst. 1981. Feeding of *Mysis relicta* Lovén on macrozooplankton. *Rept. Inst. Freshw. Res. Drottingholm* 59:75–80.

Lasenby, D. C., and R. R. Langford. 1972. Growth, life history and respiration of *Mysis relicta* in an arctic and temperate lake. *J. Fish. Res. Board Can.* 29:1701–1708.

Lasenby, D. C., and R. R. Langford. 1973. Feeding and assimilation of *Mysis relicta*. *Limnol. Oceanogr.* 18:280–285.

Latimer, D. L., A. S. Brooks, and A. M. Beeton. 1975. Toxicity of 30-minute exposures of residual chlorine to copepods *Limnocalanus macrurus* and *Cyclops biscuspidatus thomasi*. *J. Fish. Res. Board Can.* 32(12):2495–2501.

Lewkowicz, M. 1974. The communities of zooplankton in fish ponds. *Acta Hydrobiol.* 16(2):139–172.

Light, S. F. 1938. New subgenera and species of diaptomid Copepoda from the inland waters of California and Nevada. *Univ. Calif. Pub. Zool.* 43:67–78.

Light, S. F. 1939. New American subgenera of *Diaptomus* Westwood (Copepoda, Calanoida). *Trans. Amer. Microsc. Soc.* 58(4):473–484.

Lilljeborg, W. 1860. Crustacea; Cladocera. *Ofu. Ak. Forh.* 17:265.

Lim, R. P., and C. H. Fernando. 1978. Production of Cladocera inhabiting the vegetated littoral of Pinehurst Lake, Ontario, Canada. *Int. Ver. Theor. Angew. Limnol. Verh.* 20(1):225–231.

Linn, J. D., and T. C. Frantz. 1965. Introduction of the opossum shrimp (*Mysis relicta* Lovén) into California and Nevada. *Calif. Fish Game* 51:48–51.

Main, R. A. 1962. The life history and food relations of *Epischura lacustris* Forbes (Copepoda; Calanoida). Ph.D. thesis, Univ. of Michigan, Ann Arbor.

Makarewicz, J. C., and G. E. Likens. 1975. Niche analysis of a zooplankton community. *Science* (Wash. D.C.) 190:1000–1003.

Marsh, C. D. 1893. On the Cyclopidae and Calanidae of central Wisconsin. *Trans. Wis. Acad. Sci., Arts, Lett.* 9:189–224.

Marsh, C. D. 1895. On the Cyclopidae and Calanidae of Lake St. Clair, Lake Michigan, and certain of the inland lakes of Michigan. *Bull. Mich. Fish Comm.* 5:24.

Marsh, C. D. 1897. On the limnetic Crustacea of Green Lake. *Trans. Wis. Acad. Sci., Arts, Lett.* 11:179–224.

Marsh, C. D. 1899. On some points in the structure of the larvae of *Epischura lacustris* Forbes. *Trans. Wis. Acad. Sci., Arts, Lett.* 12(2):544–548.

Marsh, C. D. 1910. A revision of the North American species of *Cyclops*. *Trans. Wis. Acad. Sci., Arts, Lett.* 16:1067–1135,

Marsh, C. D. 1929. Distribution and key of the North American copepods of the genus *Diaptomus*, with a description of a new species. *Proc. U.S. Natl. Mus.* 75:1–27.

Marsh, C. D. 1933. Synopsis of the calanoid crustaceans, exclusive of the Diaptomidae, found in fresh and brackish waters chiefly of North America. *Proc. U.S. Natl. Mus.* 82(18):1–58.

Marzolf, G. R. 1965a. Vertical migration of *Pontoporeia affinis* (Amphipoda) in Lake Michigan. In *Proc. 8th Conf. Great Lakes Res.*, pp. 133–140. Univ. Mich. Great Lakes Res. Div. Publ. 13.

Marzolf, G. R. 1965b. Substrate relations of the burrowing amphipod *Pontoporeia affinis* in Lake Michigan. *Ecology* 46(5):579–592.

Maschwitz, D. E., R. J. Wedlund, and H. J. Wiegner. 1976. *Minnesota, Lake Superior water quality study: a survey of the nearshore waters in the Duluth area and near the mouths of the Gooseberry and Cascade rivers.* Minnesota Pollution Control Agency, Roseville, Minn.

McClaren, I. A. 1963. Effects of temperature on growth of zooplankton and the adaptive value of vertical migration. *J. Fish. Res. Board Can.* 20(3):685–727.

McClaren, I. A. 1969. Temperature adaptation of copepod eggs from the arctic to the tropics. *Biol. Bull.* (Woods Hole) 137(3):486–493.

McNaught, D. C. 1978. Spatial heterogeneity and niche differentiation in zooplankton of Lake Huron. *Int. Ver. Theor. Angew. Limnol. Verh.* 20(1):341–346.

McNaught, D. C., and M. Buzzard. 1973. Changes in zooplankton populations in Lake Ontario 1939–1972. In *Proc. 16th Conf. Great Lakes Res.* pp. 76–86. International Association Great Lakes Research.

McNaught, D. C., M. Buzzard, D. Griesmer, and M. Kennedy. 1980. *Zooplankton grazing and population dynamics relative to water quality in southern Lake Huron.* U.S. Environmental Protection Agency, Ecol. Res. Ser. EPA-600/3-80-069.

McNaught, D. C., and A. D. Hasler. 1966. Photoenvironemnts of planktonic Crustacea in Lake Michigan. *Int. Ver. Theor. Angew. Limnol. Verh.* 16:194–203.

McQueen, D. J. 1969. Reduction of zooplankton standing stocks by predaceous *Cyclops biscuspidatus thomasi* in Marion Lake, British Colombia. *J. Fish. Res. Board Can.* 26(6):1605–1618.

McQueen, D. J. 1970. Grazing rates and food selection in *Diaptomus oregonensis* from Marion Lake, British Columbia. *J. Fish. Res. Board Can.* 17(1):13–20.

McWilliam, P. S. 1970. Seasonal changes in abundance and reproduction in the "opossum shrimp," *Mysis relicta* Lovén, in Lake Michigan. Master's thesis, Univ. of Sydney, Australia.

Medeira, P. T., A. S. Brooks, and D. B. Seale. 1982. Excretion of total phosphorus, dissolved reactive phosphorus, ammonia, and urea by Lake Michigan *Mysis relicta*. *Hydrobiologia* 93:145–154.

Moore, J. E. 1952. The Entomostraca of southern Saskatchewan. *Can. J. Zool.* 30:410–450.

Moore, J. W. 1979a. Ecology of subarctic populations of *Cyclops bicuspidatus thomasi* Forbes and *Diaptomus ashlandi* Marsh (Copepoda). *Crustaceana* (Leiden) 36(3):237–248.

Moore, J. W. 1979b. Ecology of a subarctic population of *Pontoporeia affinis* Lindstrom (Amphipoda). *Crustaceana* (Leiden) 36(3):267–276.

Mordukhai-Boltoskoi, E. D. 1958. Preliminary notes on the feeding of carnivorous cladocerans *Leptodora* and *Bythotrephes*. *Dokl. Biol. Sci.* 122:828–830.

Mordukhai-Boltoskoi, E. D. 1965. On the males and gamogenetic females of the Caspian Polyphemidae (Cladocera). *Crustaceana* (Leiden) 12(2):113–123.

Morgan, M. D., and A. M. Beeton. 1978. Life history and abundance of *Mysis relicta* in Lake Michigan. *J. Fish. Res. Board Can.* 35(9):1165–1170.

Morsell, J. W., and C. R. Norden. 1968. Food habits of the alewife *Alosa pseudoharengus* (Wilson) in Lake Michigan. In *Proc. 11th Conf. Great Lakes Res.*, pp. 96–102. International Association Great Lakes Research.

Mozley, S. C., and R. P. Howmiller 1977. *Environmental status of the Lake Michigan region vol. 6: zoobenthos of Lake Michigan.* National Technical Information Service ANL/ES-40.

Norden, C. R. 1968. Morphology and food habits of the larval alewife, *Alosa pseudoharengus* (Wilson) in Lake Michigan. In *Proc. 11th Conf. Great Lakes Res.*, pp. 103–110. International Association Great Lakes Research.

O'Brien W. J., and F. DeNoyelles, Jr. 1974. Filtering rate of *Ceriodaphnia reticulata* in pond waters of varying phytoplankton concentrations. *Amer. Midl. Nat.* 91(2):509–512.

Olson, T. A., and T. O. Odlaug. 1966. Limnological observations on western Lake Superior. In *Proc. 9th Conf. Great Lakes Res.*, pp. 109–118. Univ. Mich. Great Lakes Res. Div. Publ. 15.

Paquette, M., and B. Pinel-Alloul. 1982. Cycles de développement de *Skistodiaptomus oregonensis*, *Tropocyclops prasinus* et *Cyclops scutifer* dans la zone limnétique du lac Cromwell, Saint-Hippolyte, Quebec. *Can. J. Zool.* 60:139–151.

Parker, J. I. 1980. Predation by *Mysis relicta* on *Pontoporeia hoyi*: a food chain link of potential importance in the Great Lakes. *J. Great Lakes Res.* 6(2):164–166.

Patalas, K. 1969. Composition and horizontal distribution of crustacean plankton in Lake Ontario. *J. Fish. Res. Board Can.* 26(8):2135–2164.

Patalas, K. 1971. Crustacean plankton communities in lakes in the Experimental Lakes Area, northwestern Ontario. *J. Fish. Res. Board Can.* 28(2):231–244.

Patalas, K. 1972. Crustacean plankton and the eutrophication of St. Lawrence Great Lakes. *J. Fish. Res. Board Can.* 29(10):1451–1462.

Patt, D. I. 1947. Some cytological observations of the Nähroboden of *Polyphemus pediculus* Linneus. *Trans. Amer. Microsc. Soc.* 66(4):344–353.

Pearse, A. S. 1921. Distribution and food of the fishes of Green Lake, Wisconsin in summer. *Bull. U.S. Bureau Fisheries* 37:254–272.

Pejler, B. 1965. Regional-ecological studies of Swedish freshwater zooplankton. *Zool. Bidrag Uppsala* 36:407–515.

Pennak, R. W. 1944. Diurnal movements of zooplankton organisms in some Colorado mountian lakes. *Ecology* 25(4):387–403.

Pennak, R. W. 1949. Annual limnological cycles in some Colorado reservoir lakes. *Ecol. Monogr.* 19(1):233–267.

Pennak, R. W. 1978. *Fresh-water invertebrates of the United States.* 2nd ed. John Wiley and Sons, New York.

Poppe, S. A. 1880. Über eine neue Art der Gattung *Temora* Baird. *Abh. Naturwiss Ver. Bremen* 7:55–60.

Porter, K. G. 1977. The plant-animal interface in freshwater ecosystems. *Amer. Sci.* (65):159–170.

Powers, C. F., and W. P. Alley. 1967. Some preliminary observations on the depth distribution of macrobenthos in Lake Michigan. In *Studies on the Environment and Eutrophication of Lake Michigan*, J. C. Ayers and D. C. Chandler, eds., pp. 112–125. Univ. Mich. Great Lakes Res. Div. Spec. Rep. 30.

Price, J. W. 1958. Cryptic speciation in the *vernalis* group of Cyclopidae. *Can. J. Zool.* 36:285–303.

Price, J. W. 1963. A study of the food habits of some Lake Erie fish. *Bull. Ohio Biol. Surv.* 2(1):1–89.

Pritchard, A. L. 1929. The alewife (*Pomolobus pseudoharengus*) in Lake Ontario. *Univ. Toronto Stud. Biol. 33; Publ. Ontario Fish. Res. Lab.* 39:39–54.

Pritchard, A. L. 1931. Taxonomic and life history studies of the ciscoes of Lake Ontario. *Univ. of Toronto Stud. Biol. No. 35* and *Publ. Ontario Fish. Res. Lab.* 41:1–78.

Putnam, H. D. 1963. A study of the nutrients, productivity, and plankton in western Lake Superior. Ph.D. thesis, University of Minnesota, Minneapolis.

Reed, E. B. 1963. Records of freshwater Crustacea from arctic and subarctic Canada. *Can. Natl. Mus. Bull.* 199:29–59.

Reighard, J. E. 1894. A biological examination of Lake St. Clair. *Bull. Mich. Fish. Comm.* 4:1–60.

Reynolds, J. B., and G. M. Degraeve. 1972. Seasonal population characteristics of the opossum shrimp *Mysis relicta* in southeastern Lake Michigan 1970–1971. In *Proc. 15th Conf. Great Lakes Res.*, pp. 117–121. International Association Great Lakes Research.

Richard, J. 1895. Revision des Cladocères. *Ann. Sci. Nat. Zool. Ser VII* 18:279–389.

Richard, J. 1896. Revision des Cladocères, duxieme partie. *Ann. Sci. Nat. Zool. Ser. VIII* 2:187–360.

Richman, S. 1964. Energy transformation studies on *Diaptomus oregonensis*. *Int. Ver. Theor. Angew. Limnol. Verh.* 15(2):654–659.

Richman, S. 1966. The effect of phytoplankton concentration on the feeding rate of *Diaptomus oregonensis*. *Int. Ver. Theor. Angew. Limnol. Verh.* 16(1):392–398.

Richman, S., S. A. Bohon, and S. E. Robins. 1980. Grazing interactions among freshwater calanoid copepods. In *Evolution and ecology of zooplankton communities*, W. C. Kerfoot, ed., pp. 219–233. U. Press of New England, Hanover, New Hampshire.

Ricker, K. E. 1959. The origin of two glacial relict crustaceans in North America, as related to Pleistocene glaciation. *Can. J. Zool.* 37:871–893.

Rieman, B. E., and C. M. Falter. 1981. Effects of the establishment of *Mysis relicta* on the macrozooplankton of a large lake. *Trans. Amer. Fish. Soc.* 110(5):613–620.

Riessen, H. P., and W. J. O'Brien. 1980. Re-evaluation of the taxonomy of *Daphnia longiremis* Sars 1862 (Cladocera): description of a new morph from Alaska. *Crustaceana* (Leiden) 38(1):1–11.

Rigler, F. H. 1972. The Char Lake project: a study of energy flow in a high arctic lake. In *Productivity problems of freshwater*, Z. Kajak and A. Hillbricht-Ilkowska, eds., pp 287–300. Polish Scientific Publishers.

Rigler, F. H., and J. M. Cooley. 1974. The use of field data to derive population statistics of multivoltine copepods. *Limnol. Oceanogr.* 19(4):636–655.

Rigler, F. H. and R. R. Langford. 1967. Congeneric occurrences of species of *Diaptomus* in southern Ontario lakes. *Can. J. Zool.* 45(1):81–90.

Robertson, A. 1966. The distribution of calanoid copepods in the Great Lakes, In *Proc. 9th Conf. Great Lakes Res.*, pp. 129–139. Univ. Mich. Great Lakes Res. Div. Publ. 15.

Robertson, A., and W. P. Alley. 1966. Comparative study of Lake Michigan macrobenthos. *Limnol. Oceanogr.* 11:576–583.

Robertson, A., and J. E. Gannon. 1981. Annotated checklist of the free-living copepods of the Great Lakes. *J. Great Lakes Res.* 7(4):382–393.

Robertson, A., C. Gehrs, B. Hardin, and G. Hunt. 1974. *Culturing and ecology of Diaptomus clavipes and Cyclops vernalis*. U.S. Environmental Protection Agency, Ecol. Res. Ser. EPA-660/3-74-006.

Robertson, A., C. F. Powers, and R. F. Anderson. 1968. Direct observations on *Mysis relicta* from a submarine. *Limnol. Oceanogr.* 13:700–702.

Roff, J. C. 1972. Aspects of the reproductive biology of the planktonic copepod *Limnocalanus macrurus* Sars. *Crustaceana* (Leiden) 22:155–160.

Roff, J. C. 1973. Oxygen consumption of *Limnocalanus macrurus* Sars (Calanoida, Copepoda) in relation to environmental conditions. *Can. J. Zool.* 51(8):877–885.

Roff, J. C., and J. C. H. Carter. 1972. Life cycle and seasonal abundance of the copepod *Limnocalanus macrurus* Sars in a high arctic lake. *Limnol. Oceanogr.* 17(3):363–370.

Rolan, R. G., N. Zack, and M. Pritschau. 1973. Zooplankton Crustacea of the Cleveland nearshore area of Lake Erie, 1971–1972. In *Proc. 16th Conf. Great Lakes Res.*, pp. 116–131. International Association Great Lakes Research.

Roth, J. C. 1973. Benton Harbor power plant limnological studies. In *Cook Plant preparational studies 1972*, J. C. Ayers and E. Siebil, eds. Univ. Mich. Great Lakes Res. Div. Part XIII Spec. Rep. 44.

Roth, J. C., and J. A. Stewart. 1973. Nearshore zooplankton of southeastern Lake Michigan, 1972. In *Proc. 16th Conf. Great Lakes Res.*, pp. 132–142. International Association Great Lakes Research.

Sandercock, G. A. 1967. A study of selected mechanisms for the coexistence of *Diaptomus spp.* in Clarke Lake, Ontario. *Limnol. Oceanogr.* 12(1):97–112.

Sars, G. O. 1862. Oversigt af de af ham: Omegnen af Christiana iagttagne Crustacea cladocera. *Forhand. Vidensk.-Selsk. Christiania* (1861):144–167, 250–302.

Sars, G. O. 1873. Om en dimorph uduikling samt generationsvexel hos *Leptodora*. *Ofvers. Dan. Selsk.*, pp. 1–15.

Sars, G. O. 1895. On some South African *Entomostraca* raised from dried mud. *Skr. Vidensk. Christiania* 8:1–56.

Sars, G. O. 1903. *An account of the Crustacea of Norway. IV. Copepoda, Calanoida*. Bergen Press, Bergen.

Sars, G. O. 1915. Entomostraca of Georgian Bay. *Contrib. Can. Biol. 1911–1914*, II:221–222.

Sars, G. O. 1918. *Crustacea of Norway. Vol. VI. Copepoda, Cyclopoida*. Bergen Museum, Bergen.

Schacht, F. W. 1897. The North American species of *Diaptomus*. *Bull. Ill. Nat. Hist.* 5:97–207.

Schacht, F. W. 1898. North American centropagidae belonging to the genera *Osphranticum*, *Limnocalanus*, and *Epischura*. *Bull. Ill. Nat. Hist.* 5:225–270.

Scheffer, V. B. and R. J. Robinson. 1939. A limnological study of Lake Washington. *Ecol. Monogr.* 9(1):95–143.

Schindler, D. W., and B. Noven. 1971. Vertical distribution and seasonal abundance of zooplankton in two shallow lakes of the Experimental Lakes Area, northwestern Ontario. *J. Fish. Res. Board Can.* 28(2):245–256.

Schneberger, E. 1937. The biological and economic importance of the smelt in Green Bay. *Trans. Amer. Fish.Soc.* 66:139–142.

Schumacher, R. 1966. Successful introduction of *Mysis relicta* Lovén into a Minnesota lake. *Trans. Amer. Fish. Soc.* 95(2):216.

Scourfield, D. J. and J. P. Harding. 1941. A key to the British species of freshwater Cladocera with notes on their ecology. *Freshw. Biol. Assoc. of the British Empire* 5:1–55.

Seale, D. B., and M. E. Boraas. 1982. Influence of experimental conditions on nitrogenous excretion by Lake Michigan *Mysis relicta* (Lovén): laboratory studies with animals acclimated in *Fragillaria*. *Hydrobiologia* 93:163–170.

Segerstråle, S. G. 1937. Studien über die Bodentierwelt in südfinnländischen Küstengewässern III. Zur Morphologie und Biologie des Amphipoden *Pontoporeia affinis*, nebst einer Revision der Pontoporeia-Systematik. *Comm. Biol. Helsingfors* 7(1):1–183.

Segerstråle, S. G. 1962. The immigration and prehistory of the glacial relicts of Eurasia and North America. A survey and discussion of modern views. *Int. Rev. Gesamten Hydrobiol.* 37(1):1–25.

Segerstråle, S. G. 1967. Observations of summer breeding in populations of the glacial relict *Pontoporeia affinis* Lindstrom (Crustacea, Amphipoda) living at greater depths in the Baltic Sea with notes on the reproduction of *Pontoporeia femorata* Kroyer. *J. Exp. Mar. Bio. Ecol.* 1(1):55–64.

Segerstråle, S. G. 1971a. The distribution and morphology of *Pontoporeia affinis* Lindstrom. f. *brevicornis* (Crustacea, Amphipoda) inhabiting North American lakes, with a description of a new aberrant male form from the area. *Comm. Biol.* 38:1–19.

Segerstråle, S. G. 1971b. On summer-breeding in populations of *Pontoporeia affinis* (Crustacea, Amphipoda) living in lakes of North America. *Commentat. Biol. Soc. Sci. Fenn.* 44:1–18.

Segerstråle, S. G. 1971c. Light control of the reproductive cycle of *Pontoporeia affinis* Lindstrom (Crustacea, Amphipoda). *J. Exp. Mar. Biol. Ecol.* 5(3):272–275.

Segerstråle, S. G. 1972. The zoogeographic problems involved in the presence of glacial relict *Pontoporeia affinis* (Crustacea, Amphipoda) in Lake Washington USA. *J. Fish. Res. Board Can.* 28(9):1331–1334.

Selgeby, J. H. 1974. *Littoral crustacean zooplankton of the Apostle Islands region of Lake Superior, May–December, 1971.* Great Lakes Fish. Lab. Admin. Rep. U.S. Fish and Wildlife Service, Ashland, Wisconsin.

Selgeby, J. H. 1975a. Life histories and abundance of crustacean zooplankton in the outlet of Lake Superior, 1971–1972. *J. Fish. Res. Board Can.* 32:461–470.

Selgeby, J. H. 1975b. *Composition of the crustacean zooplankton of Lake Superior with an annotated bibliography of zooplankton research on Lake Superior.* Great Lakes Fish. Lab. Admin. Rep. U.S. Fish and Wildlife Service; Ashland, Wisconsin.

Sell, D. W. 1982. Size frequency estimates of secondary production by *Mysis relicta* in Lakes Michigan and Huron. *Hydrobiologia* 93:69–78.

Shrivastava, H. 1974. *Macrobenthos of Lake Huron.* J. Fish. Res. Board Can. Tech. Rept. 44.

Sibley, C. K. 1929. The food of certain fishes of the Lake Erie drainage basin, In *A biological survey of the Erie-Niagara system*, pp. 180–188. Suppl. 18th An. Rept. (1928), New York State Conservation Department.

Siefken, M., and K. B. Armitage. 1968. Seasonal variation in metabolism and organic nutrients in three *Diaptomus* (Crustacea, Copepoda). *Comp. Biochem. Physiol.* 24(2):591–609.

Sierszen, M. E., and A. S. Brooks. 1982. The release of dissolved organic carbon as a result of diatom fragmentation during feeding by *Mysis relicta*. *Hydrobiologia* 93:155–162.

Smirnov, N. N. 1962. *Eurycercus lamellatus* (O. F. Müller) (Chydoridac, Cladocera): field observations and nutrition. *Hydrobiologia* 20:280–294.

Smirnov, N. N. 1966a. The taxonomic significance of the trunk limbs of the Chydoridae (Cladocera). *Hydrobiologia* 27:337–343.

Smirnov, N. N. 1966b. *Alonopsis* (Chydoridae, Cladocera): morphology and taxonomic position. *Hydrobiologia* 27:113–136.

Smirnov, N. N. 1966c. Morphology of Chydoridae (Cladocera) and their distribution. *Int. Ver. Theor. Angew. Limnol. Verh.* 16(3):1673–1676.

Smirnov, N. N. 1969. *Allonella* and *Dunhevedia* (Chydoridae, Cladocera): morphology of trunk limbs. *Hydrobiologia* 33:547–560.

Smith, K., and C. H. Fernando. 1978. A guide to the freshwater calanoid and cyclopoid copepod Crustacea of Ontario. Univ. of Waterloo Biol. Ser. 18.

Smith, L. L., Jr., and J. B. Moyle. 1944. A biological survey and fishery management plan for the streams of the Lake Superior north shore watershed. *Minn. Dept. Conserv. Tech. Bull.* No. 1:1–228.

Smith, S. I. 1871. The fauna of Lake Superior at great depths. *Amer. Nat.* 5:722.

Smith, S. I. 1874a. The Crustacea of the fresh waters of the United States. *Rep. U.S. Comm. Fish and Fisheries* (1872–1873) Pt. 2:637–665.

Smith, S. I. 1874b. Sketch of the invertebrate fauna of Lake Superior. *Rep. U.S. Comm. Fish and Fisheries* (1872–1873) Pt. 2:690–707.

Smith, W. E. 1970. Tolerance of *Mysis relicta* to thermal shock and light. *Trans. Amer. Fish. Soc.* 99(2):418–422.

Smith, W. E. 1972. Culture, reproduction and temperature tolerance of *Pontoporeia affinis* in the laboratory. *Trans. Amer. Fish. Soc.* 101(2):253–256.

Smyly, W. J. P. 1958. Distribution and seasonal abundance of Entomostraca in moorland parts near Windermere. *Hydrobiologia* 11:59–72.

Smyly, W. J. P. 1961a. The life cycle of the freshwater copepod *Cyclops leuckarti* Claus in Esthwaite Water. *J. Anim. Ecol.* 30:153–169.

Smyly, W. J. P. 1961b. Some aspects of the biology of *Cyclops leuckarti*. *Int. Ver. Theor. Angew. Limnol. Ver.* 14:946–949.

Stenson, J. A. E. 1973. On predation and *Holopedium gibberum* (Zaddach) distribution. *Limnol. Oceanogr.* 18(6):1005–1010.

Stenson, J. A. E. 1976. Effect of predator influence on composition of *Bosmina* spp. populations. *Limnol. Oceanogr.* 21(6):814–822.

Stewart, J. A. 1974. Lake Michigan zooplankton communities in the area of the Cook nuclear plant. In *The biological, chemical, and physical character of Lake Michigan in the vicinity of the Donald C. Cook nuclear plant.* ed. E. Seible and J. C. Ayers, eds., pp. 211–232. Univ. Mich. Great Lakes Res. Div., Spec. Rep. 51.

Stimpson, W. 1871. On the deep-water fauna of Lake Michigan. *Amer. Nat.* 4:403–405.

Stringer, G. E. 1967. Introduction of *Mysis relicta* Lovén into Kalamalka and Pinaus Lakes, British Columbia. *J. Fish. Res. Board Can.* 24(2):463–465.

Strøm, K. M. 1946. The ecological niche. *Nature* 157:375.

Swain, W. R., R. W. Magnuson, J. D. Johnson, T. A. Olson, and T. O. Odlaug. 1970a. Vertical migration of zooplankton in western Lake Superior. In *Proc. 13th Conf. Great Lakes Res.*, pp. 619–639. International Association Great Lakes Research.

Swain, W. R., T. A. Olson, and T. O. Odlaug. 1970b. The ecology of the second trophic level in Lakes Superior, Michigan, and Huron. *Univ. Minn. Water Resource Ctr. Bull.* 26:1–151.

Tash, J. C. 1971. The zooplankton of fresh and brackish waters of the Cape Thompson area, northern Alaska. *Hydrobiologia* 38(1):93–121.

Tattersall, M. W., and O. S. Tattersall. 1951. *The British Mysidacea*. The Ray Society, London.

Teter, H. E. 1960. The bottom fauna of Lake Huron. *Trans. Amer. Fish. Soc.* 89:193–197.

Thienemann, A. 1926. *Holopedium gibberum* in Holstein. *Z. Morph. Oekol. Tiere* 5:755–776.

Thomas, M. L. H. 1966. Benthos of four Lake Superior bays. *Can. Field. Nat.* 80:200–212.

Thomas, M. P. 1963. Notes on the presence of *Sida crystallina* in the plankton and the origin of the freshwater plankton. *Arch. Hydrobiol.* 59(1):103–109.

Torke, B. G. 1974. *An illustrated guide to the identification of the planktonic crustacea of Lake Michigan with notes on their ecology.* Univ. of Wisc.-Milw. Ctr. for Great Lakes Stud., Spec. Rep. 17.

Torke, B. G. 1975. The population dynamics and life histories of crustacean zooplankton at a deep-water station in Lake Michigan. Ph.D. thesis, University of Wisconsin.

Torke, B. G. 1976. *A key to the identification of the cyclopoid copepods of Wisconsin, with notes on their distribution and ecology.* Wis. Dept. Nat. Resour. Rep. 88.

Tressler W. L., T. S. Austin, and E. Orban. 1953. Seasonal variation of some limnological factors in Irondequoit Bay, New York. *Amer. Midl. Nat.*49:873–903.

Upper Lakes Reference Group. 1977. *The waters of Lake Huron and Lake Superior. Vol. III, Part B. Lake Superior.* International Joint Commission, Windsor, Ontario.

Vanderploeg, H. A., J. A. Bowers, O. Chapelski, and H. K. Soo. 1982. Measuring *in situ* predation by *Mysis relicta* and observations on underdispersed microdistributions of zooplankton. *Hydrobiologia* 93:109–120.

Van Oosten, J., and H. L. Deason. 1938. The food of the lake trout (*Christivomer namaycush namaycush*) and of the lawyer (*Lota maculosa*) of Lake Michigan *Trans. Amer. Fish. Soc.* 67:155–177.

Vorce, C. M. 1881. Forms observed in water of Lake Erie. *Proc. Amer. Microsc. Soc.* 4:51–60.

Vorce, C. M. 1882. Microscopic forms observed in the waters of Lake Erie. *Trans. Amer. Microsc. Soc.* 5:187–196.

Wagner, N. 1870. *Hyalosoma dux,* a new form of amphipod Crustacea. Transactions of the first meeting of Russian naturalists at St. Petersburg, 1868, pp. 218–238. (In Russian).

Ward, E. B. 1940. A seasonal population study of Entomostraca in the Cincinnati region. *Amer. Midl. Nat.* 23(3):635–691.

Warren, E. 1901. A preliminary account of the development of the free-swimming nauplius of *Leptodora hyalina* (Lillj.). *Proc. R. Soc. London., B. Biol. Sci.* 68:210–218.

Watson, N. H. F. 1976. Seasonal distribution and abundance of crustacean zooplankton in Lake Erie, 1970. *J. Fish. Res. Board Can.* 33(3):612–621.

Watson, N H. F., and G. F. Carpenter. 1974. Seasonal abundance of crustacean zooplankton and net plankton biomass of Lakes Huron, Erie, and Ontario. *J. Fish. Res. Board Can.* 31(3):309–317.

Watson, N. H. F., and B. M. Smallman. 1971. The role of photoperiod and temperature in the induction and termination of an arrested development in two species of freshwater cyclopid copepods. *Can. J. Zool.* 49(6):855–862.

Watson, N. H. F., and J. B. Wilson. 1978. Crustacean zooplankton of Lake Superior. *J. Great Lakes Res.* 4(3–4):481–496.

Welch, P. S. 1935. *Limnology* (1st Ed.). McGraw-Hill Book Co., New York.

Wells, L. 1960. Seasonal abundance and vertical movements of planktonic crustacea in Lake Michigan. *U.S. Fish Wildlf. Serv. Fish. Bull.* 60:343–369.

Wells, L. 1968. Daytime distribution of *Pontoporeia affinis* off the bottom of Lake Michigan. *Limnol. Oceanogr.* 13:703–705.

Wells, L. 1970. Effects of alewife predation on zooplankton populations in Lake Michigan. *Limnol. Oceanogr.* 15:556–565.

Wells, L., and A. M. Beeton. 1963. Food of the bloater, *Coregonus hoyi,* in Lake Michigan. *Trans. Amer. Fish. Soc.* 92(3):245–255.

Westwood, J. O. 1836. Cyclops. *Partington's Cyclopaedia.*

Whiteside, M. C. 1970. Danish chydorid Cladocera: modern ecology and core studies. *Ecol. Monogr.* 40(1):79–118.

Wickliff, E. L. 1920. Food of young small-mouth black bass in Lake Erie. *Trans. Amer. Fish. Soc.* 50:364–371.

Wierzbicka, M. 1953. *Limnocalanus macrurus* (G. O. Sars) u jeziorach pólnocnawschodniej Polski. *Fragm. faun. Mus. Zool. Polonici* 6(20):525–540.

Wiley, A. 1920. Report on marine copepoda collected during the Canadian arctic expedition. *Rep. Can. Arct. Exped.* 1913–1918 7,K:1–46.

Williams, J. B., and M. C. Whiteside. 1978. Population regulation of the Chydoridae in Lake Itasca, Minnesota. *Int. Ver. Theor. Angew. Limnol. Verh.* 20(4):2484–2489.

Wilson, C. B. 1960. *The macroplankton of Lake Erie.* pp. 145–1720. U.S. Fish. Wildlf. Serv., Spec. Sci. Rep. 334.

Wilson, J. B., and J. C. Roff. 1973. Seasonal vertical distributions and diurnal migration patterns of Lake Ontario crustacean zooplankton. In *Proc. 16th Conf. Great Lakes Res.,* pp. 190–203. International Association Great Lakes Research.

Wilson, M. S. 1959. Free-living Copepoda: Calanoida. In *Freshwater Biology,* 2nd ed. W. T. Edmondson, ed., pp. 738–794. J. Wiley and Sons, New York.

Wilson, M. S., and J. C. Tash. 1966. The euryhaline copepod genus *Eurytemora* in fresh and brackish waters of the Cape Thompson region, Chukchi Sea, Alaska. *Proc. U.S. Natl. Mus.* 118:553–576.

Wilson, M. S., and H. C. Yeatman. 1959a. Free-living Copepoda. In *Fresh-water Biology,* 2nd ed. W. T. Edmondson, ed., pp. 735–738. J. Wiley and Sons, New York.

Wilson, M. S., and H. C. Yeatman. 1959b. Free-living Copepoda: Harpacticoida. In *Fresh-water Biology,* 2nd ed. W. T. Edmondson, ed., pp. 815–861. J. Wiley and Sons, New York.

Winner, R. W., and J. F. Haney. 1967. Spatial and seasonal distribution of planktonic Cladocera in a small reservoir. *Ohio J. Sci.* 67:274–288.

Wolfert, D. R. 1965. Food of young-of-the-year walleyes in Lake Erie. *U.S. Fish Wildlf. Serv. Bull.* 65(2):489–494.

Woltereck, R. 1932. Races, associations, and stratifications of pelagic daphnids, in some lakes of Wisconsin and other regions of the U.S. and Canada. *Trans. Wis. Acad. Sci., Arts, Lett.* 27:487–521.

Wong, C. K. 1981. Predatory feeding behavior of *Epischura lacustris* and prey defense. *Can. J. Fish. Aquat. Sci.* 38(3):275–279.

Wright, S. 1955. *Limnological survey of western Lake Erie.* U.S. Fish Wildlf. Serv. Spec. Sci. Rep. Fish. 139:1–341.

Yeatman, H. D. 1944. American cyclopoid copepods of the *vir-*

idis-vernalis group (including a description of *Cyclops car-olinianus* n.sp.). *Amer. Midl. Nat.* 32(1):1–90.

Yeatman, H. C. 1959. Free-living Copepoda: Cyclopoida. In *Fresh-water Biology*, 2nd ed., W. T. Edmondson, ed., pp. 793–815. J. Wiley and Sons, New York.

Zaddach, E. G. 1855. *Holopedium gibberum* ein neues Crusta-cean aus der Familie Branchiopoden. *Arch. Naturgesch* 21:159–188.

Zhdanova, G. A. 1969. Comparative characteristics of the life cycle and productivity of *Bosmina longirostris* O. F. Müller and *Bosmina coregoni* Baird in the Kiev reservoir. *Hydrobiol. J.* 5:8–15.

Plate 1 *Leptodora kindti*, ♀

Plate 2 *Leptodora kindti,* ♂

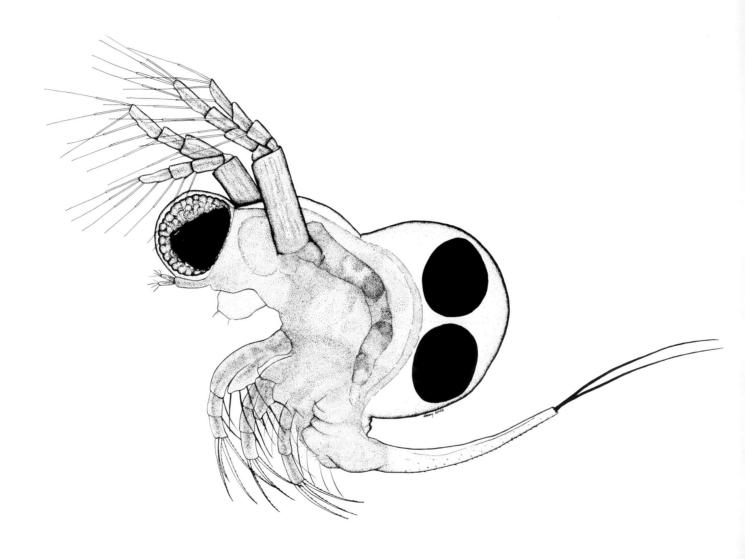

Plate 3   *Polyphemus pediculus*, ♀

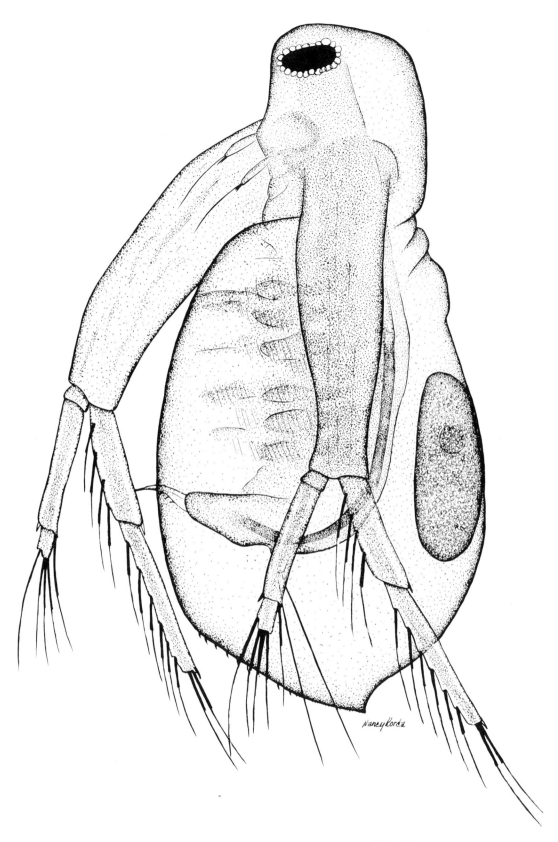

Plate 4 *Diaphanosoma birgei*, ♀, lateral view

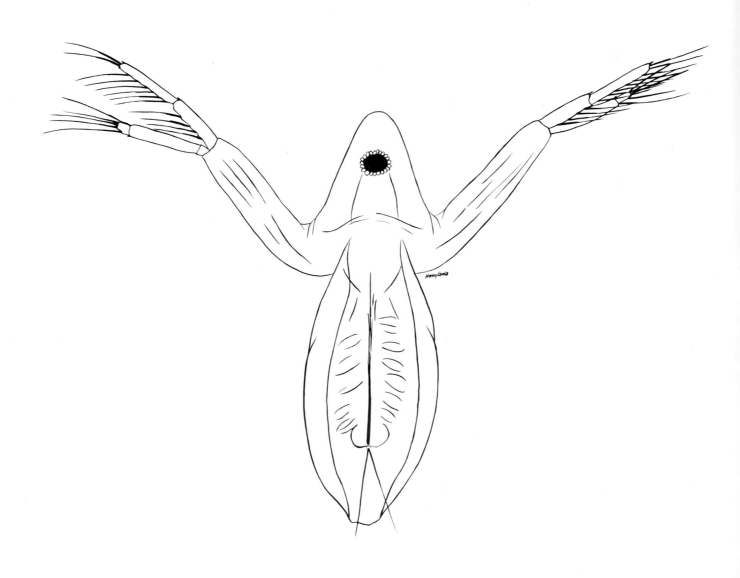

Plate 5   *Diaphanosoma birgei*, ♀, ventral view

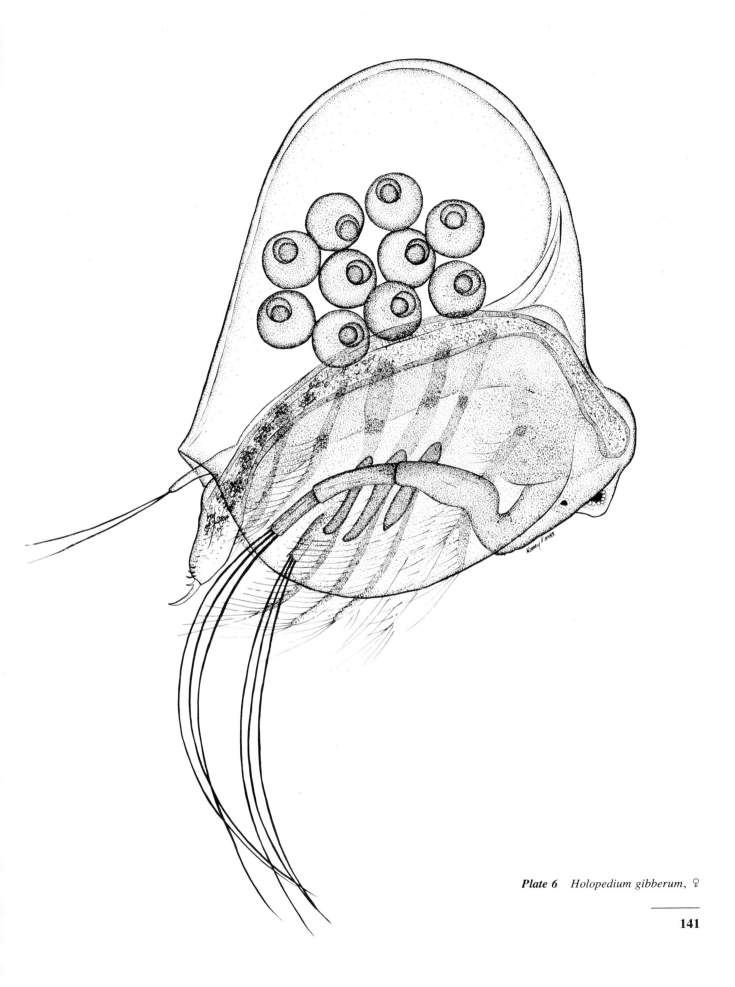

Plate 6 *Holopedium gibberum*, ♀

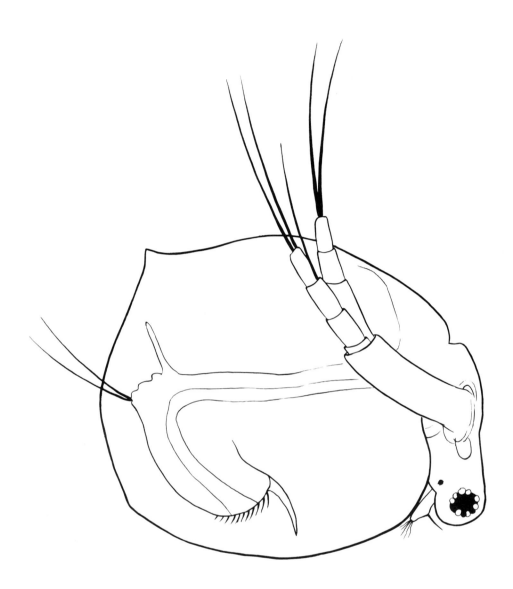

Plate 7 Ceriodaphnia, ♀

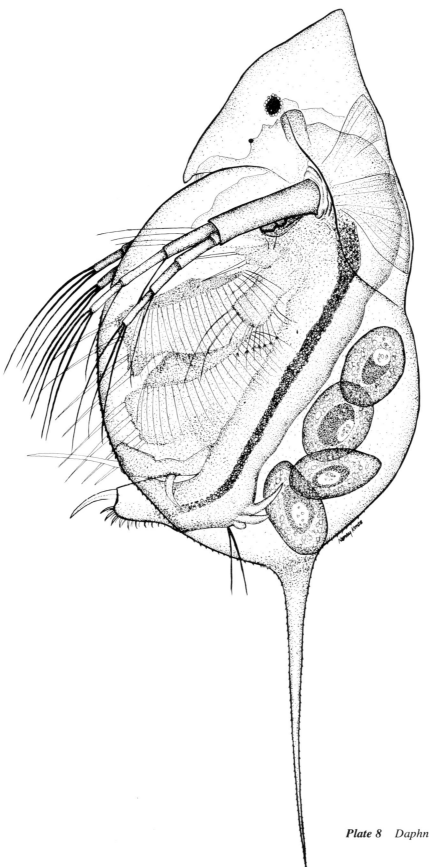

Plate 8   *Daphnia galeata mendotae*, ♀

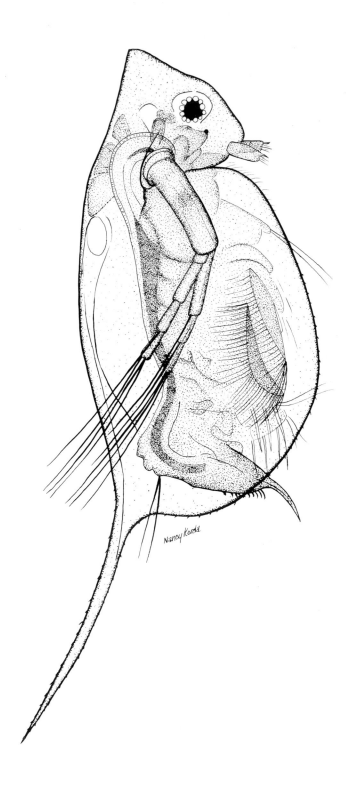

Plate 9 *Daphnia galeata mendotae*, ♂

144

Plate 10   Variations in head shape of *Daphnia galeata mendotae*

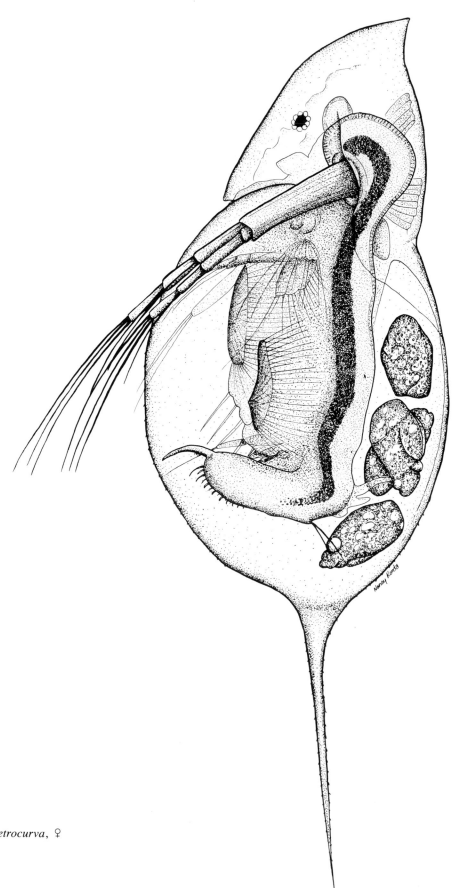

Plate 11 *Daphnia retrocurva*, ♀

146

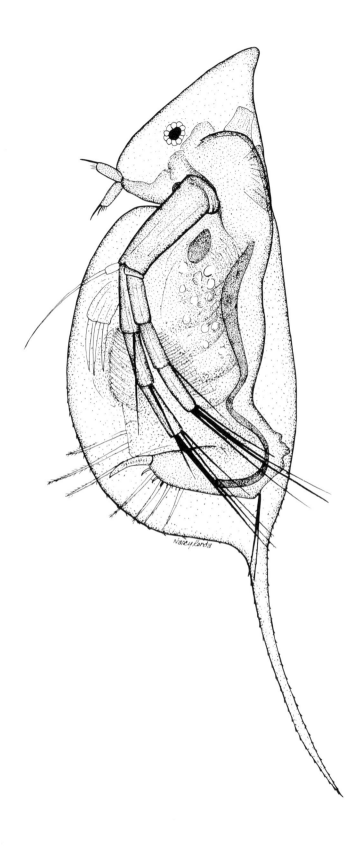

Plate 12   *Daphnia retrocurva*, ♂

Plate 13 Variations in head shape of *Daphnia retrocurva*, ♀ ♀

148

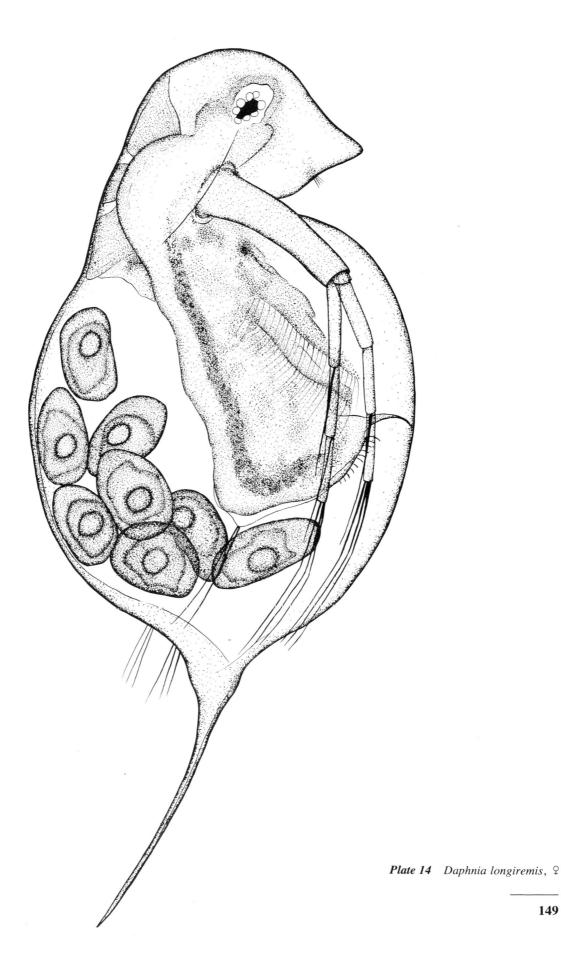

Plate 14   *Daphnia longiremis*, ♀

149

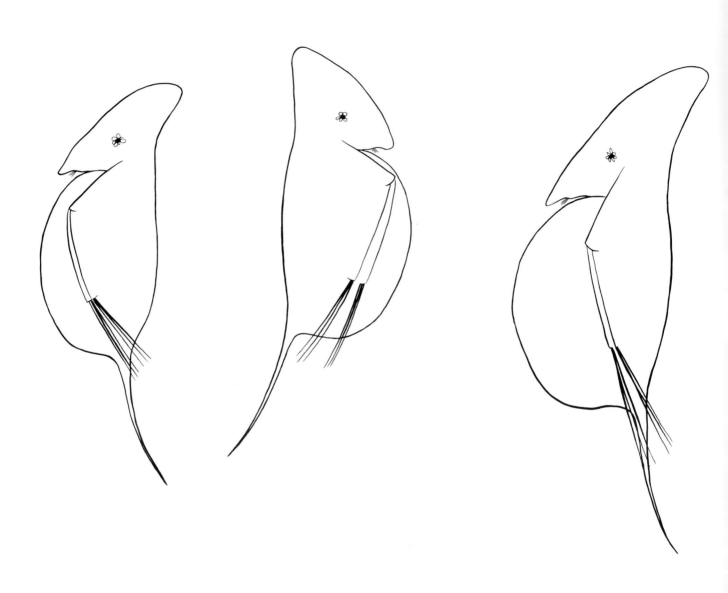

Plate 15 Variations in body form and head shape of immature *Daphnia longiremis*

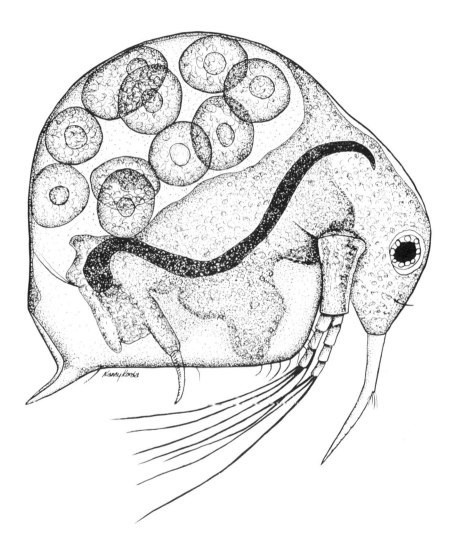

Plate 16 *Bosmina longirostris*, ♀

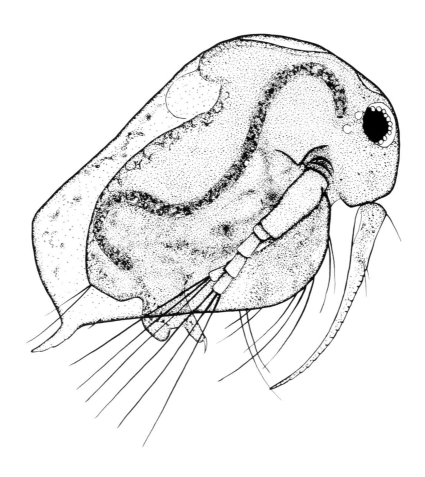

Plate 17 *Bosmina longirostris,* ♂

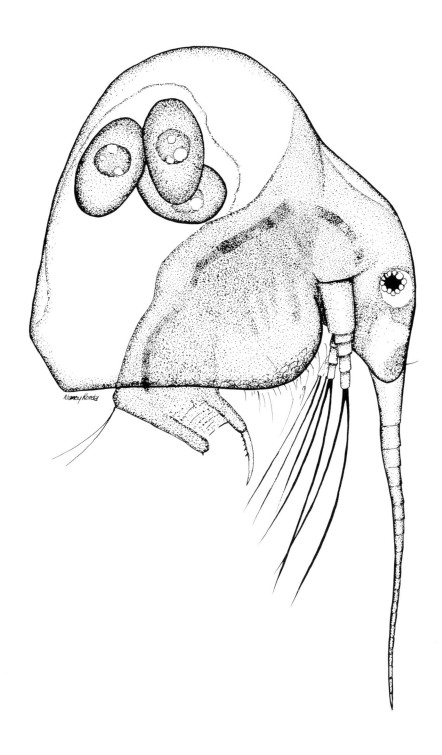

Plate 18   *Eubosmina coregoni,* ♀

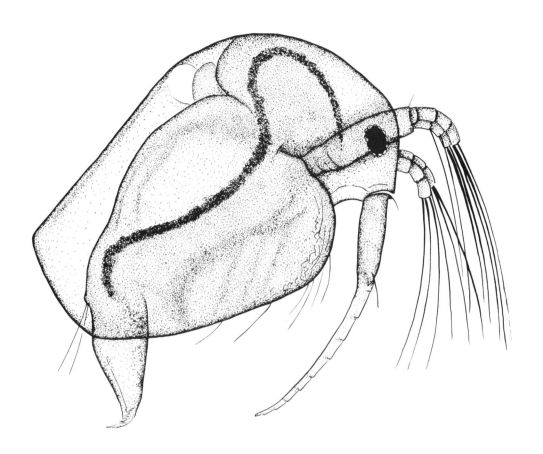

Plate 19 *Eubosmina coregoni*, ♂

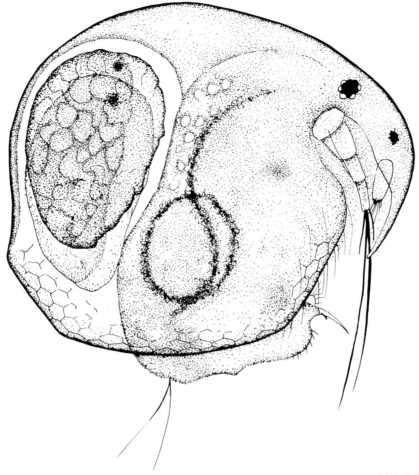

Plate 20 *Chydorus spaericus*, ♀

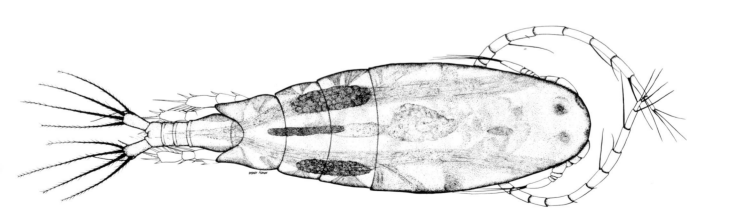

Plate 21 *Senecella calanoides*, ♀, dorsal view

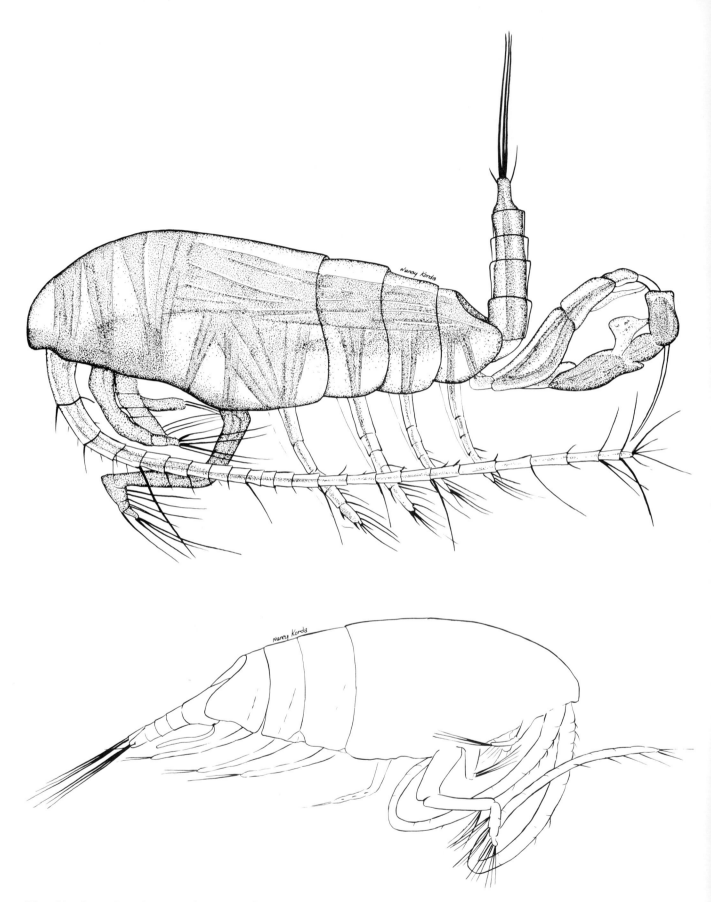

Plate 22 *Senecella calanoides*, ♂ (top) and ♀ (bottom), lateral view

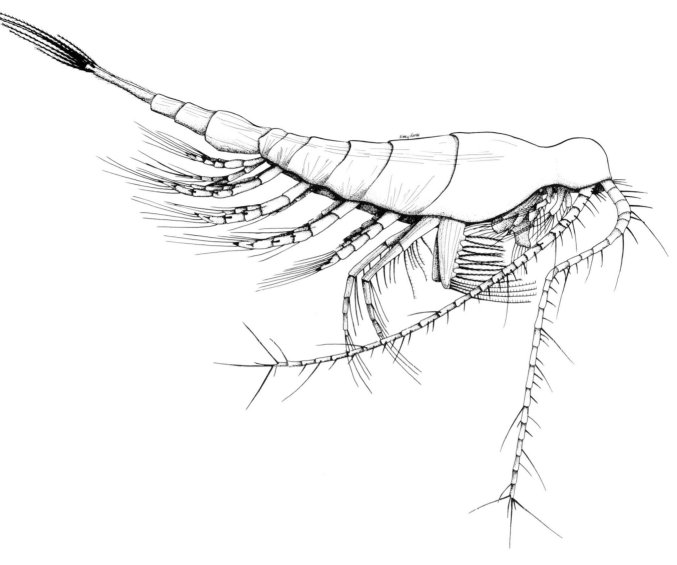

Plate 23 *Limnocalanus macrurus,* ♀

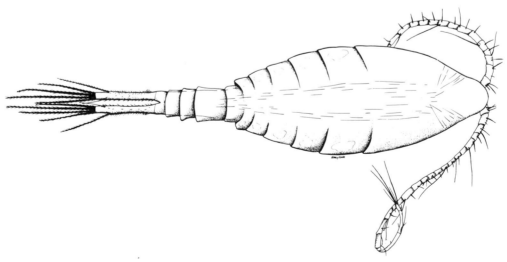

Plate 24 *Limnocalanus macrurus,* ♂

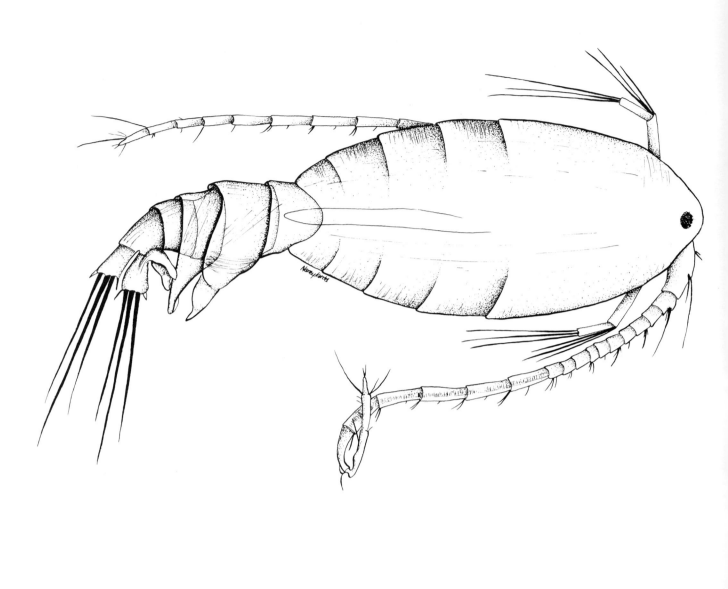

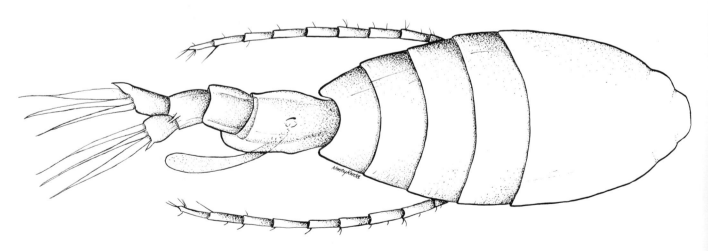

Plate 25   *Epischura lacustris*, ♂ (top) and ♀ (bottom)

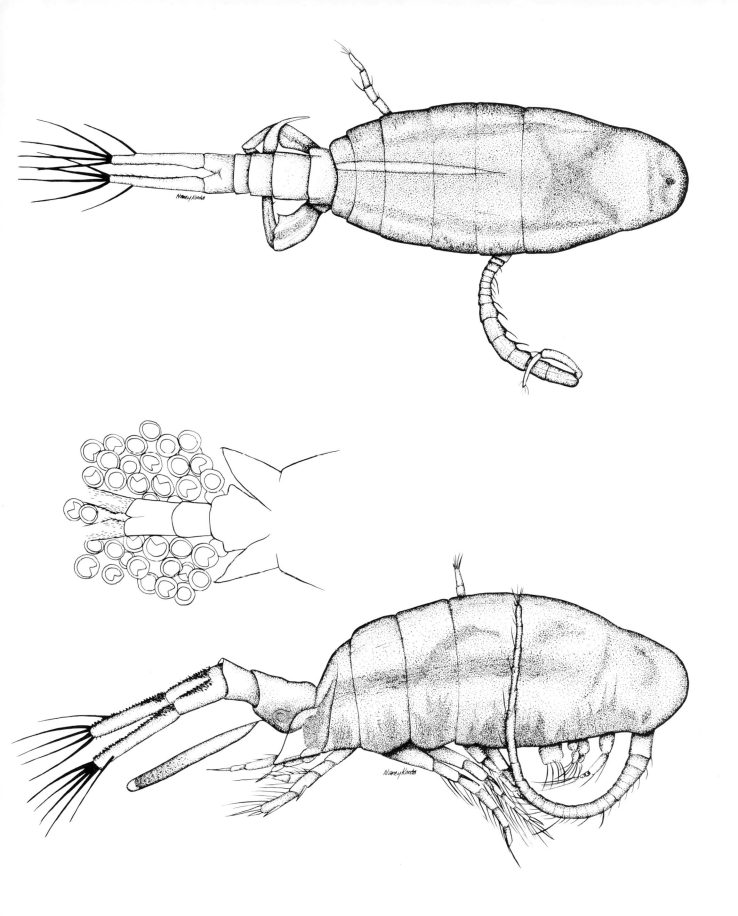

Plate 26   *Eurytemora affinis*, ♂ (top) and ♀ (bottom)

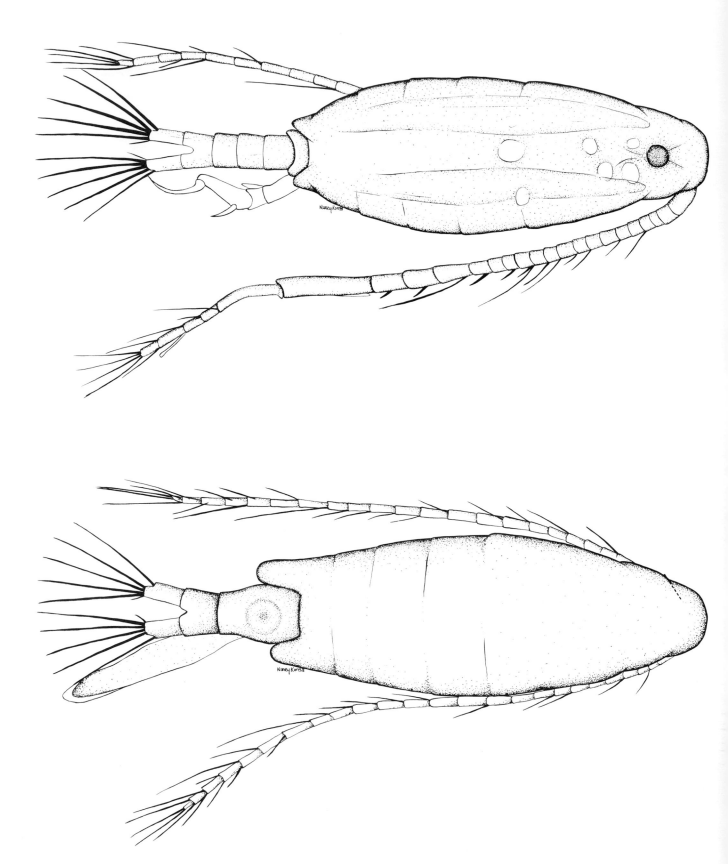

Plate 27 *Leptodiaptomus ashlandi*, ♂ (top) and ♀ (bottom)

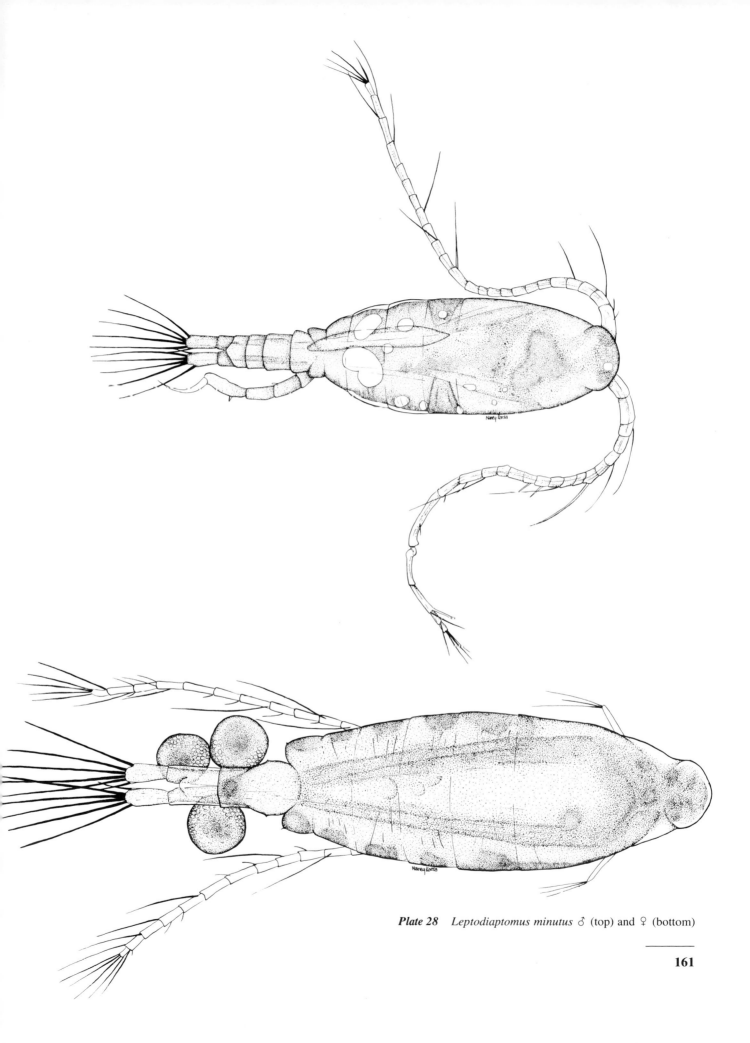

Plate 28 *Leptodiaptomus minutus* ♂ (top) and ♀ (bottom)

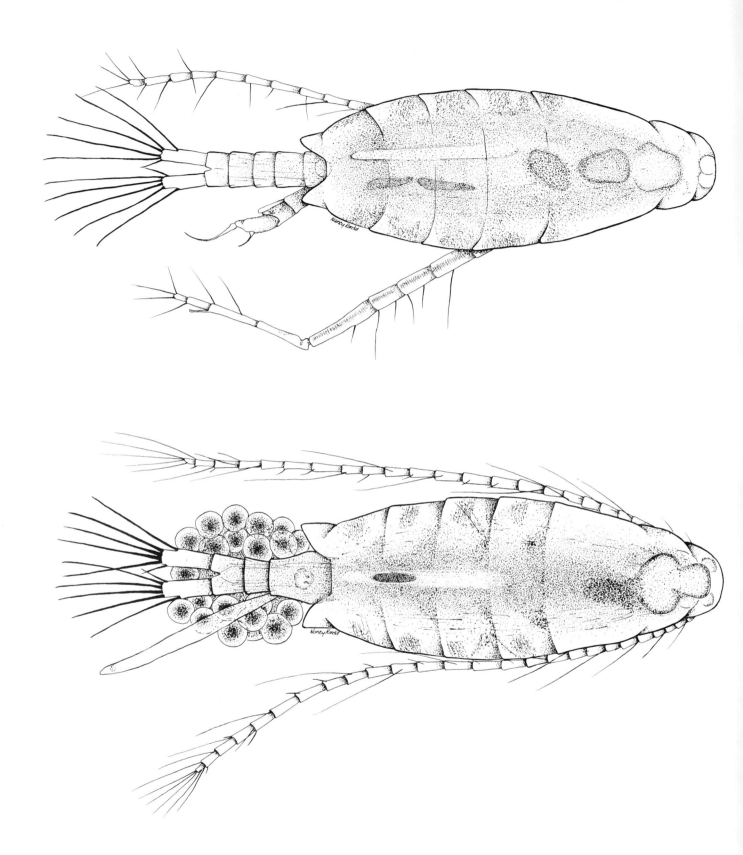

Plate 29 *Leptodiaptomus sicilis*, ♂ (top) and ♀ (bottom)

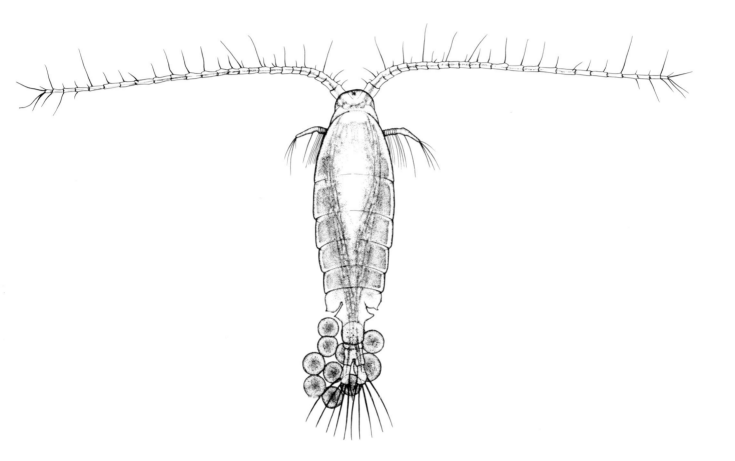

Plate 30   *Leptodiaptomus siciloides*, ♀

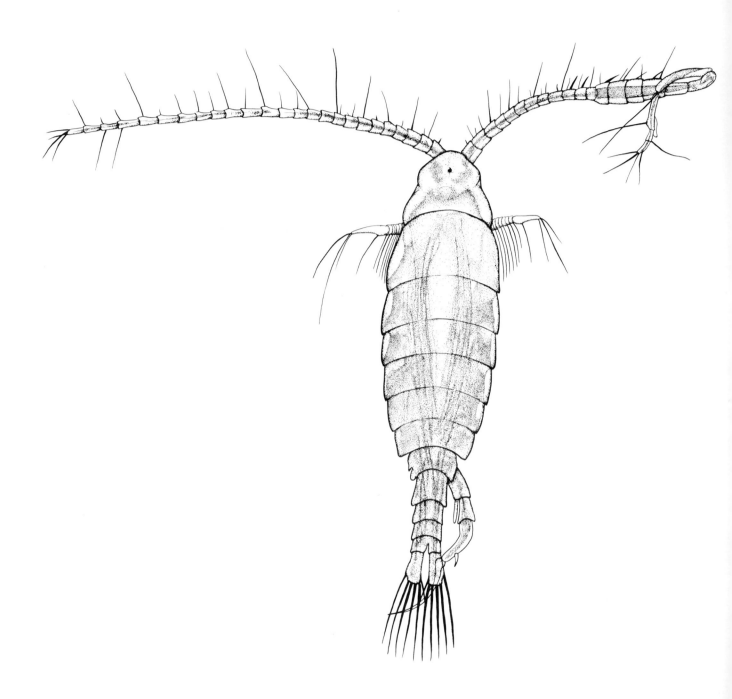

Plate 31 *Leptodiaptomus siciloides*, ♂

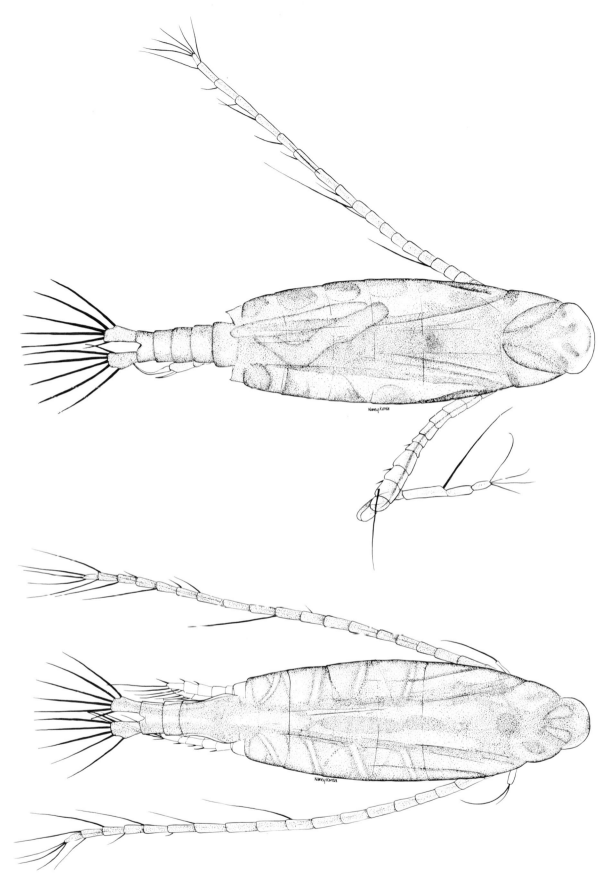

Plate 32   *Skistodiaptomus oregonensis*, ♂ (top) and ♀ (bottom)

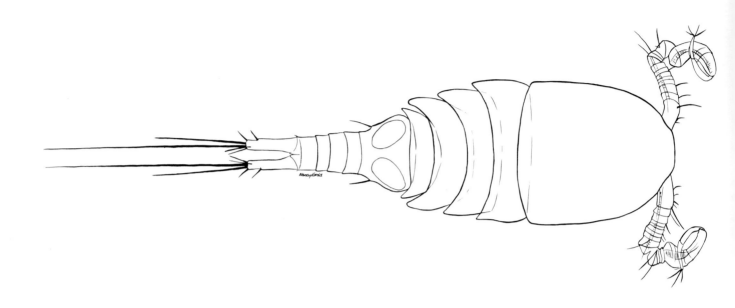

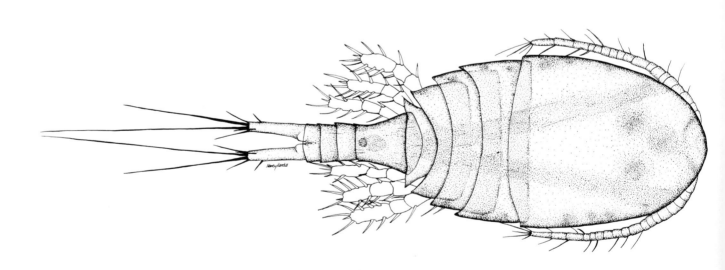

Plate 33 *Acanthocyclops vernalis*, ♂ (top) and ♀ (bottom)

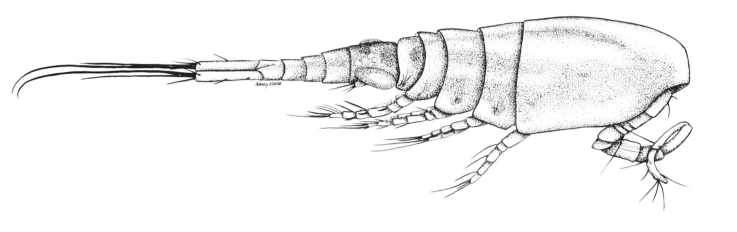

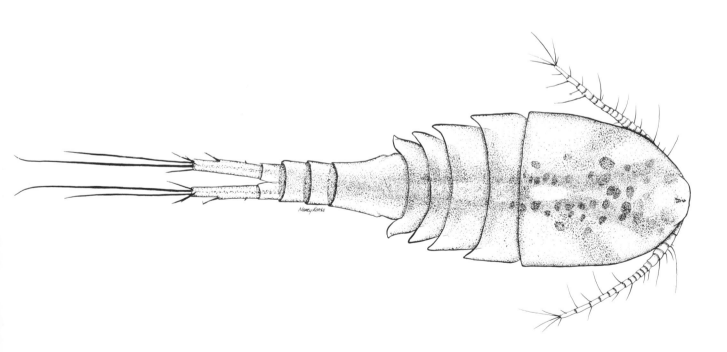

Plate 34 *Diacyclops thomasi*, ♂ (top) and ♀ (bottom)

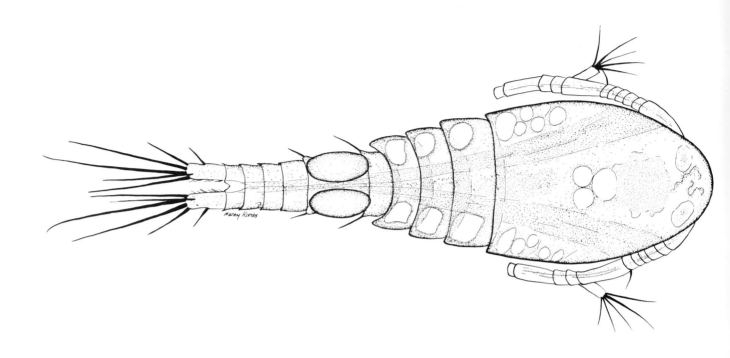

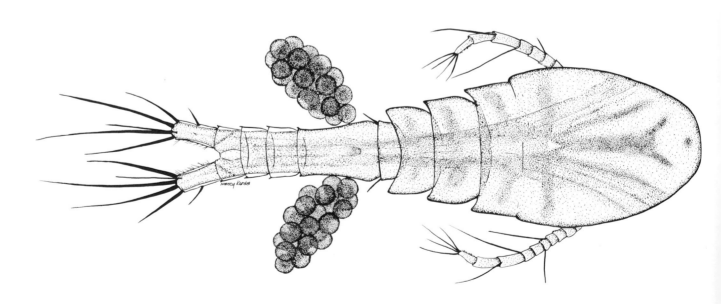

Plate 35 *Mesocyclops edax*, ♂ (top) and ♀ (bottom)

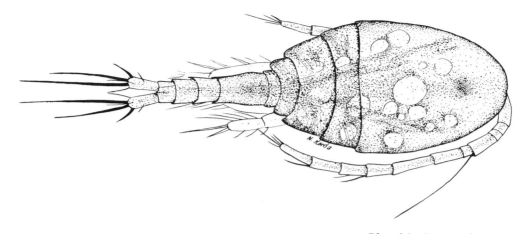

Plate 36 *Tropocyclops prasinus mexicanus*, ♀

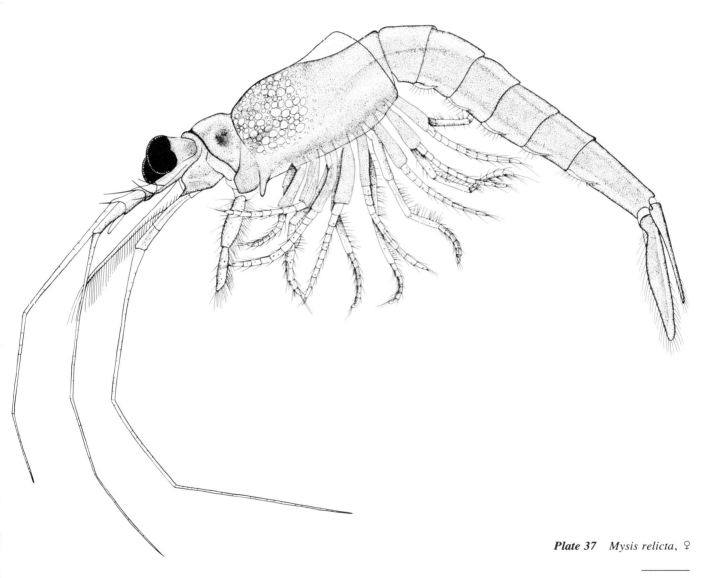

Plate 37 *Mysis relicta*, ♀

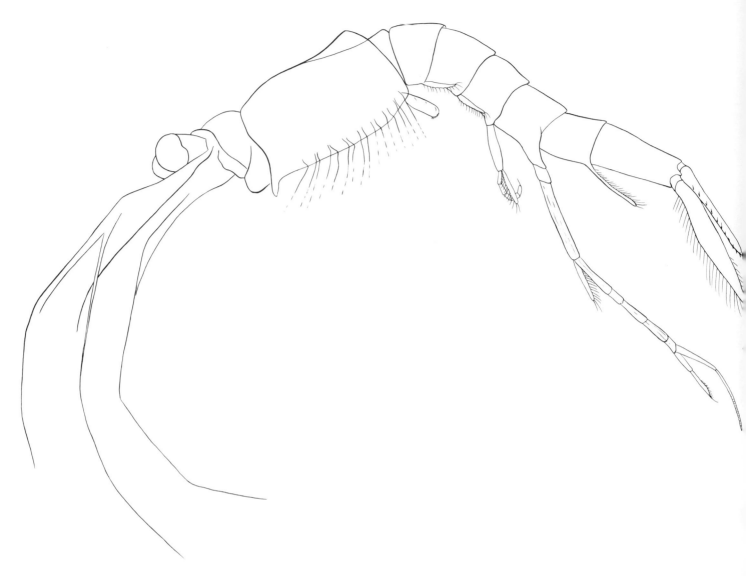

Plate 38　*Mysis relicta,* ♂

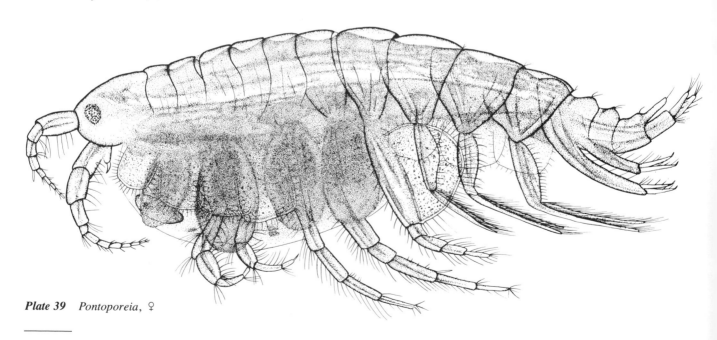

Plate 39　*Pontoporeia,* ♀

170

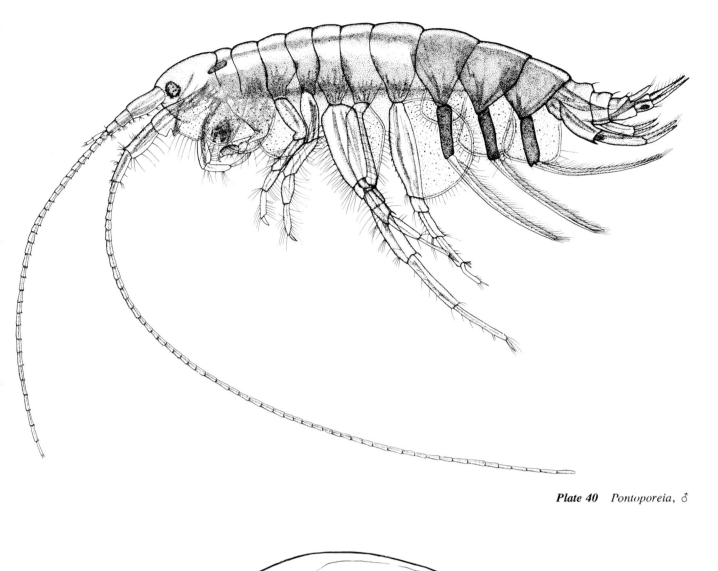

Plate 40   *Pontoporeia*, ♂

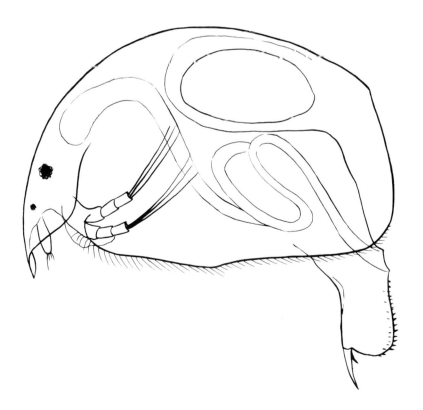

Plate 41   *Alona*, ♀

Index

Designed by Arlene Putterman
Composed by G & S Typesetters, Austin, Texas
Manufactured by Thomson-Shore, Inc., Dexter, Michigan
Text and display lines are set in Times Roman

Library of Congress Cataloging in Publication Data

Balcer, Mary D.
    Zooplankton of the Great Lakes.

    Bibliography: p.
    Includes index.
    1. Freshwater zooplankton—Great Lakes. 2. Freshwater
zooplankton—Great Lakes—Identification.   3. Crustacea—Great
Lakes.   4. Crustacea—Great Lakes—Identification. I. Korda,
Nancy L.   II. Dodson, Stanley I.   III. Title.
QL143.B35 1984   595.3′0977   83-27426
ISBN 0-299-09820-6